Work with display scree

Health and Safety (Display Screen Equipment)
Regulations 1992 as amended by the Health and Safety
(Miscellaneous Amendments) Regulations 2002

L26

GUIDANCE ON
REGULATIONS

HSE

HSE BOOKS

© Crown copyright 2003

First published 1992
Second edition 2003

ISBN 0 7176 2582 6

This guidance is issued by the Health and Safety Executive. Following the guidance is not compulsory and you are free to take other action. But if you do follow the guidance you will normally be doing enough to comply with the law. Health and safety inspectors seek to secure compliance with the law and may refer to this guidance as illustrating good practice.

Contents

Introduction

1   This book gives detailed and comprehensive guidance about work with display screen equipment (often known as VDUs). It describes how to comply with the Health and Safety (Display Screen Equipment) Regulations (referred to here as the DSE Regulations). It covers both office work and other environments where display screen equipment (DSE) may be used. It is intended for people who need to consider all the detailed implications of the law. For those who just want basic practical advice on how to comply with the Regulations when using ordinary DSE in offices, an alternative publication is our shorter, illustrated book *The law on VDUs: An easy guide*.[1]

2   This book sets out information, explanation or advice on each main part of the DSE Regulations. Throughout the document, extracts of the DSE Regulations are printed in *italics* and the guidance on them is in plain type. Where the DSE Regulations are self-explanatory no comment is offered.

3   The main risks that may arise in work with DSE are musculoskeletal disorders such as back pain or upper limb disorders (sometimes known as repetitive strain injury or RSI), visual fatigue, and mental stress. While the risks to individual users are often low (see paragraph 33), they can still be significant if good practice is not followed. DSE workers are also so numerous that the amount of ill health associated with such work is significant and tackling it is important. That is what the DSE Regulations set out to achieve. Appendix 2 gives further information on the health risks in DSE work, and explains how efforts to reduce these risks will link into the Health and Safety Commission's strategy for occupational health.

4   This book gives guidance on the Health and Safety (Display Screen Equipment) Regulations 1992, as amended by the Health and Safety (Miscellaneous Amendments) Regulations 2002. The DSE Regulations came into force in 1993 to implement a European Directive, No. 90/270/EEC of 29 May 1990, on minimum safety and health requirements for work with DSE.

5   This book, revised in 2003, takes account of the recent changes to the DSE Regulations mentioned in paragraph 4. Only regulations 3, 5 and 6 have been amended and the new versions are reproduced here. The guidance has been revised in other places, to bring it up to date with changes in technology and improvements in knowledge of risks and how to avoid them. However, the main messages about actions employers and workers should take to prevent risks have altered very little.

6   The advice in this book covers the DSE Regulations only, and references to a regulation are, unless otherwise stated, to that part of the DSE Regulations. Employers should ensure that they also comply with general duties placed on them by other health and safety legislation, such as the Health and Safety at Work etc Act 1974 (the HSW Act),[2] the Management of Health and Safety at Work Regulations 1999,[3] the Workplace (Health, Safety and Welfare) Regulations 1992 (as amended),[4] and the Provision and Use of Work Equipment Regulations 1998 (as amended).[5] There are other HSE guidance books such as *Essentials of health and safety at work*[6] that describe the requirements of this general legislation.

7   Overlaps between general and specific legislation occur where broadly applicable legislation (such as the Regulations mentioned in paragraph 6) imposes a general duty similar to a more specific one in the DSE Regulations; examples are given in items (a) and (b) below. In such cases the legal requirement is to comply with **both** the more specific and the general

duties. However, this should not give rise to any difficulty in practice. For example in display screen work:

(a)   carrying out the suitable and sufficient analysis of workstations and risk assessment required by regulation 2 of the DSE Regulations (see paragraph 37) will also satisfy the requirement in the Management of Health and Safety at Work Regulations 1999[3] for risk assessment, as far as those workstations are concerned;

(b)   ensuring that the requirements for lighting, reflections and glare in the schedule to the DSE Regulations are met (see paragraphs 28-33 of Appendix 1) will also satisfy the requirements for suitable and sufficient lighting in the Provision and Use of Work Equipment Regulations 1998[5] and the Workplace (Health, Safety and Welfare) Regulations 1992,[4] as far as the DSE workstations are concerned.

8   Where DSE is used but such use is not covered by the DSE Regulations (because the equipment is exempt or there is no defined 'user' or 'operator' - see the guidance on regulation 1), the workers concerned are still protected by the HSW Act[2] and the other general legislation described in paragraph 6. (Examples are the requirements for suitable and sufficient lighting in the Provision and Use of Work Equipment Regulations 1998[5] and the Workplace (Health, Safety and Welfare) Regulations 1992,[4] and the general requirements for risk assessment and provision of training and information in the Management of Health and Safety at Work Regulations 1999).[3] Where a display screen is in use but the DSE Regulations do not apply, the assessment of risks and measures taken to control them should take account of ergonomic factors applicable to DSE work.

Citation, commencement, interpretation and application

(1)   These Regulations may be cited as the Health and Safety (Display Screen Equipment) Regulations 1992 and shall come into force on 1 January 1993.

(2)   In these Regulations –

(a)   "display screen equipment" means any alphanumeric or graphic display screen, regardless of the display process involved;

(b)   "operator" means a self-employed person who habitually uses display screen equipment as a significant part of his normal work;

(c)   "use" means use for or in connection with work;

(d)   "user" means an employee who habitually uses display screen equipment as a significant part of his normal work; and

(e)   "workstation" means an assembly comprising –

(i)   display screen equipment (whether provided with software determining the interface between the equipment and its operator or user, a keyboard or any other input device),

(ii)   any optional accessories to the display screen equipment,

(iii)   any disk drive, telephone, modem, printer, document holder, work chair, work desk, work surface or other item peripheral to the display screen equipment, and

(iv)   the immediate work environment around the display screen equipment.

(3)   Any reference in these Regulations to –

(a)   a numbered regulation is a reference to the regulation in these Regulations so numbered; or

(b)   a numbered paragraph is a reference to the paragraph so numbered in the regulation in which the reference appears.

Which display screen equipment is covered?

9   With a few exceptions (see paragraphs 21-26), the definition of DSE at regulation 1(2)(a) covers both conventional (cathode-ray tube) display screens and other types such as liquid crystal or plasma displays used in flat-panel screens, touchscreens and other emerging technologies. Display screens mainly used to display line drawings, graphs, charts or computer-generated graphics are included, as are screens used in work with television or film pictures (this point has been clarified in a case heard before the European Court of Justice, ECJ case C-11/99). The definition is not limited to typical office situations or computer screens but also covers, for example non-electronic display systems such as microfiche. DSE used in factories and other non-office workplaces is included (subject to the exceptions in paragraphs 21-26), although in some situations such as screens used for process control or closed-circuit television (CCTV), certain requirements may not apply (see paragraphs 55-57).

Who is a display screen user or operator?

10   The DSE Regulations are for the protection of people - employees and self-employed - who habitually use DSE for the purposes of an employer's undertaking as a significant part of their normal work.

11   Regulation 1(2)(d) defines the **employees** who are covered as **users**. Regulations 2 to 7 apply to protect users, whether they are employed to work:

(a)   at their own employer's workstation;

(b)   at a workstation at home; or

(c)   at another employer's workstation.

Further guidance on homeworkers, teleworkers and agency workers is given at paragraphs 27-31.

12   The protection afforded by regulations 2, 3 and 7 also extends to **self-employed** people who work at an employer's workstation and whose use of DSE is such that they would be users if employed. They are defined in regulation 1(2)(b) as **operators** for the purposes of the DSE Regulations.

13   Employers must therefore decide which of their employees are DSE users and whether they also make use of other users (employed by other employers) or of operators. Workers who do not input or extract information by means of DSE need not be regarded as users or operators in this context - for example many of those engaged in manufacture, sales, maintenance or cleaning of DSE. Whether or not those involved in DSE work are users or operators depends on the nature and extent of their use of the equipment.

14   It is worth bearing in mind the reasoning behind these definitions. The possible risk factors associated with DSE use are mainly those leading to musculoskeletal problems, visual fatigue and stress (see Appendix 2). The likelihood of experiencing these is related mainly to the frequency, duration, intensity and pace of spells of continuous use of DSE, allied to other factors, such as the amount of discretion the person has over the extent and methods of display screen use. The combination of factors which give rise to risks makes it impossible to lay down hard-and-fast rules (for example based on set hours' usage per day or week) about who should be classified as a user or operator.

15   Where it is clear that use of DSE is more or less continuous on most days, the individuals concerned should be regarded as users or operators. This will include the majority of those whose job mainly involves, for example DSE-based data input or sales and order processing. Where use is less continuous or frequent, other factors connected with the job must be assessed. It will generally be appropriate to classify the person concerned as a user or operator if they:

(a)   normally use DSE for **continuous or near-continuous spells of an hour or more** at a time; and

(b)   use DSE in this way **more or less daily**; and

(c)   have to **transfer information quickly** to or from the DSE;

and also need to apply high levels of **attention and concentration**; or are **highly dependent** on DSE or have **little choice** about using it; or need **special training or skills** to use the DSE.

4

16   Part-time workers should be assessed using the same criteria. For example if an employee works only two days a week but spends most of that time on DSE work, that person should definitely be considered a user.

17   Not all employers may wish to analyse each job to decide whether the person doing it is a user (or operator). It is an option to simply decide that all staff who have access to DSE will be treated as users. This can save effort and allow resources to be concentrated on identifying, prioritising and remedying risks.

18   Table 1 gives some examples to illustrate factors in the decision on who is a user. (This is **not** an exhaustive list of display screen jobs.)

Table 1 Who is a display screen user?

Some examples

Definite display screen users

Word processing worker. Employed on full-time document creation and amendment. A mix of checking documents on screen, keyboard input and formatting. Some change of posture involved in collecting work, operating printer, etc. Often five hours in total on the DSE work itself, with a lunch break and at least two breaks morning and afternoon.

Secretary or typist. Uses DSE (typically a PC and printer) for word processing of reports, memos, letters from manuscript and dictation, combined with e-mail. Some variation in workload and some degree of control over scheduling throughout the day. Typically around two or three hours DSE use daily.

Data input operator. Employed full time on continuous processing of invoices. Predominantly numeric input using numeric key pad. May be subject to a system of keystroke monitoring with associated bonus payments.

News sub-editor. Makes use of DSE more or less continuously with peak workloads. Some text input to abridge/précis stories, but mainly scanning copy for fact, punctuation, grammar and size.

Journalist. Pattern of work may be variable but includes substantial use of DSE. Information collected by field or telephone interviews (which may involve use of a portable computer) followed by, typically, several hours' text input while working on a story. Work likely to be characterised by deadlines and interruptions. Some days may contain periods of less intense work but with more prolonged keyboard text entry and composition.

Telesales/customer complaints/accounts enquiry/directory enquiry operator. Often employed in a call centre, and making fairly constant use of DSE while making or taking telephone calls to or from customers. Amount of keyboard work may be low, but reliant on DSE to do the job.

Microelectronics assembly or testing operative. Uses DSE to view tests or completed components. Uses the DSE frequently for, typically, repetitive operations throughout the working shift. A dedicated test panel may be used in place of a standard keyboard.

Television editing technician. Main work is using DSE to view and process video sequences to prepare them for broadcasting.

Security control room operative. Main work is to monitor a bank of display screens showing the pictures from CCTV cameras, and operate controls to select, zoom in, etc on particular images.

Air traffic controller. Main task is monitoring of purpose-designed screens for air traffic movements combined with communication with air crew on navigation, etc. High visual and mental workload and consequences of errors critical. Shift-work.

Financial dealer. Uses a dedicated workstation typically with multiple display screens. Variable and unpredictable workload. Often highly stressful situations with information overload. Often long hours.

Graphic designer. Works on multimedia applications. Intensive scrutiny of images at high resolution. Large screens. Page make-up. Multiple input devices. Colour systems critical.

Librarian. Carries out intensive text input to databases; accessing and checking on database records; creating summaries and reports; combining data already on the system with new material inputted. DSE work either intensive throughout the day on most days, or more intermittent but still forming around half of the total working time.

Possible display screen users - depending on the circumstances

The following are examples of jobs whose occupants may or may not be designated as display screen users, depending on circumstances. In reaching a decision, employers will need to judge the relative importance of different aspects of the work, weighing these against the factors discussed in paragraph 15 of the main guidance and bearing in mind the risks to which the job holder may be exposed. If there is doubt over whether an individual is a user, carrying out a risk assessment (see regulation 2) should help in reaching a decision.

Scientist/technical adviser. Works on DSE for word processing of a few letters/memos per day; and monitoring of e-mail for a short period, average around half an hour in total, on most days. At irregular intervals, uses DSE intensively for data analysis of research results.

Discussion: This scientist's daily use of DSE is relatively brief, non-intense and he or she would have a good deal of discretion over when and how the equipment was used. Judged against this daily use, he or she would not be classed as a user. However this decision might be reversed if the periods of use for analysis of research results were at all frequent, of long duration and intensive. And increasingly, DSE is being used for a widening range of tasks (in the laboratory, as in many other workplaces), making it more likely that the workers concerned will be users.

Client manager. Working in a large management accounting consultancy. Dedicated DSE on desk. Daily use of e-mail to read, prepare and send messages. Typically 1.5 to 2 hours daily.

Discussion: Whether or not such managers are classed as users will depend on the extent and nature of their use of e-mail. For example how continuous is use of the screen and/or keyboard during each period of use? Is there discretion as to the extent of use of e-mail? How long is the total daily use of e-mail, and how long is the DSE used for anything else?

Bank or building society customer support officer. Has shared use of office, desk and display screen workstation. DSE used during interviews with clients to interrogate HQ database to obtain customer details, transactions, etc and to prepare financial proposals and estimates.

Discussion: Decision will be influenced by: what proportion of the individual worker's time is spent using the DSE; are there any prolonged and/or intensive periods of use; and what are the consequences of errors (this factor may be relevant if the job involves inputting financial data as well as searching a database).

Airline check-in clerk. The workload in the job as a whole varies throughout the day, with occasional peaks of intensive work associated with particular flights. Use of DSE follows a predictable pattern; typically, used as part of most transactions but may not be a significant proportion of total working time.

Discussion: There needs to be consideration of how equipment is used and for what purpose. Is the DSE used during most parts of the check-in process or only a few of them? Is the workload of transactions high? What proportion of each transaction involves viewing the screen or keying in data? To what extent is interaction with the screen rushed and intensive? What are the consequences of errors?

Community care worker. Uses a portable computer to make notes during and/or following interviews or visits in the field.

Discussion: Decisions on whether or not those using laptops are users need to be made on the same basis as if they were using non-portable equipment. Some of the specific minimum requirements in the Schedule may not be applicable to portables in prolonged use, as the inherent characteristics of the task (for example the need for mobility and portability) may rule them out. However, it is important that such work is properly assessed, that users are adequately trained, and that measures are taken to control risks. More guidance on portables is given in Appendix 3.

Receptionist. Job involves frequent use of DSE, for example to check or enter details of each visitor and/or provide them with information.

Discussion: The nature, frequency and duration of periods of DSE work need to be assessed. Some, perhaps most, receptionists would not be classed as users if most of their work consists of face-to-face contact and/or phone calls, with a display screen only being used occasionally – (see final part of this table, below).

Definitely not display screen users

Senior manager. Working in a large organisation, using DSE for occasional monitoring of state of markets or other data, or more frequent but brief enquiries. Does not depend on DSE for most of their work and can choose whether or not to use it personally.

Senior manager. Uses DSE (sometimes a laptop) infrequently for short periods, for generation/manipulation of statistics for presentation at meetings.

Receptionist. If work is mainly concerned with customer-public interaction, with the possibility of using DSE occasionally for limited purposes, for example obtaining details of staff telephone numbers, locations; or intermittent monitoring of a single CCTV screen.

19   Table 2 shows how the criteria in paragraph 15 relate to the job examples in Table 1.

What is a 'workstation'?

20   Under regulation 1(2)(e), a workstation exists wherever there is DSE (including portable DSE in prolonged use, see paragraph 23). For all DSE the workstation, as defined, is the assembly including the screen, keyboard, other parts of the computer and its accessories (such as the mouse or other input device), the desk, chair and the immediate work environment. Some of these items are specifically mentioned in the DSE Regulations, but anything else in the immediate work environment is also part of the workstation.

Application

> *(4)   Nothing in these Regulations shall apply to or in relation to –*
>
> *(a)   drivers' cabs or control cabs for vehicles or machinery;*
>
> *(b)   display screen equipment on board a means of transport;*
>
> *(c)   display screen equipment mainly intended for public operation;*
>
> *(d)   portable systems not in prolonged use;*
>
> *(e)   calculators, cash registers or any equipment having a small data or measurement display required for direct use of the equipment; or*
>
> *(f)   window typewriters.*

21   Where any of the exclusions in regulation 1(4) apply, none of the duties imposed by the DSE Regulations will apply to or in connection with the use of the equipment that is excluded. However, the proviso at paragraph 8 applies here too. Employers should still ensure that, so far as is reasonably practicable, the health and safety of those using the equipment are not put at risk. The general duties on employers and others under the HSW Act, and other general health and safety legislation (see paragraphs 6-8), are still applicable and particular attention should be paid to ergonomics in this context. Ergonomics[7] is the science of making sure that work tasks, equipment, information and the working environment are suitable for every worker, so that work can be done safely and productively. Ergonomic factors relevant to DSE work are discussed further in Appendix 1.

22   The exclusion in regulation 1(4)(c) is for DSE mainly intended for public operation, such as cashpoint machines, and microfiche readers and computer terminals in public libraries, etc. 'Public operation' means operation by anyone who is not an employee or a self-employed person, hence the DSE Regulations do not apply to workstations provided for school pupils or students. (It is nevertheless good practice for students and pupils to be trained to set up and use DSE and workstations in a way that minimises the risks. Further guidance for schools/colleges about health and safety with information and communications technology has been produced by BECTA, see Appendix 6). The exclusion in regulation 1(4)(c) does not extend to DSE available for operation by the public but mainly intended for users or operators.

23   Regulation 1(4)(d) excludes only portable DSE that is not in prolonged use. So the DSE Regulations do apply to portable DSE in prolonged use – which can include laptop and handheld computers, personal digital assistant

Table 2 Deciding who is a 'user' (or 'operator') under the DSE Regulations. How the criteria in the guidance apply to the job examples in Table 1

Job example	Does the jobholder's DSE work involve:							Decision
	Continuous spells of an hour or more?	Daily use of DSE?	Fast information transfer?	High attention and concentration?	High dependency on the DSE?	Little choice whether or not to use the DSE?	Special training or skills?	
Word processing	Yes	Yes	Yes	Maybe	Yes	Yes	Yes	Definitely 'users' or 'operators'
Secretary	Yes	Yes	Yes	Maybe	Maybe	Maybe	Yes	
Data input operator	Yes	Yes	Yes	Maybe	Yes	Yes	Yes	
News sub-editor	Yes	Yes	Yes	Maybe	Yes	Yes	Yes	
Journalist	Yes	Yes	Yes	Maybe	Maybe	Maybe	Yes	
Telesales/ complaints/ enquiries	Yes	Yes	Yes	Yes	Yes	Yes	Yes	
Assembly/ testing operative	Yes	Yes	Yes	Yes	Yes	Yes	Yes	
TV editing technician	Yes	Yes	Yes	Maybe	Yes	Yes	Yes	
CCTV control room worker	Yes	Yes	Maybe	Maybe	Yes	Yes	Maybe	
Air traffic controller	Yes	Yes	Yes	Yes	Yes	Yes	Yes	
Financial dealer	Yes	Yes	Yes	Yes	Yes	Yes	Yes	
Graphic designer	Yes	Yes	Yes	Maybe	Yes	Yes	Yes	
Librarian	Yes	Yes	Yes	Maybe	Yes	Yes	Yes	
Scientist/ technical adviser	Maybe	Yes	Maybe	Maybe	Maybe	Yes	No	May be 'users' or 'operators'
Client manager	Maybe	Yes	Maybe	Maybe	Maybe	Yes	No	
Banking customer support	Maybe	Yes	Maybe	Maybe	Yes	Yes	Maybe	
Airline check-in clerk	Maybe	Yes	Maybe	Maybe	Yes	Yes	Yes	
Community care fieldworker	Maybe	Maybe	Maybe	Maybe	Maybe	Maybe	No	
Receptionist (first example)	Maybe	Yes	Maybe	No	Maybe	Yes	Yes	
Senior manager (first example)	No	Yes	No	Maybe	Maybe	Maybe	No	Not 'users' or 'operators'
Senior manager (second example)	Maybe	No	No	Maybe	Maybe	No	No	
Receptionist (second example)	No	No	No	No	No	Maybe	No	

Yes means this does apply on a typical working day (not necessarily on all days).
Maybe means either this applies to the job on some days but not others, or that it applies to some such jobs but not others. In either case, there may be a need for further thought or investigation to reach a decision.
No means this never applies, or there are only occasional days in the year when it applies.
Note that Table 2 relates to the specific examples described in Table 1. Decisions on whether a job holder is a user or operator might be different for other jobs with the same job titles.

9

devices and some portable communications devices - but see also paragraph 25. While there are no hard-and-fast rules on what constitutes 'prolonged' use, portable equipment that is habitually in use by a DSE user for a significant part of his or her normal work, as explained in paragraph 15, should be regarded as covered by the DSE Regulations. While some of the specific minimum requirements in the Schedule may not be applicable to portables in prolonged use, employers should still ensure that such work is assessed and measures taken to control risks. Appendix 3 gives further guidance on practical steps to safeguard those using portables.

24   There is an exclusion in regulation 1(4)(e) for cash registers. This is intended to cover equipment whose function is to calculate/record money transactions at a point of sale. It is the way the equipment is used, rather than its physical characteristics, that determines whether it is covered by the exemption for cash registers. For example point-of-sale equipment that is used as a cash register but is also regularly used for other purposes would **not** be excluded from the scope of the DSE Regulations by regulation 1(4)(e). Examples of such other purposes might include the checking of seating plans when selling cinema tickets, or the calling up and examination of customer details when processing bank account transactions.

25   The exclusion in regulation 1(4)(e) for 'small data or measurement display' is there because such displays are usually not intensively monitored by workers for long continuous spells. This exclusion covers, for example much scientific and medical equipment, such as cardiac monitors, oscilloscopes, and instruments with small displays showing a series of digits. However, with the merging of information and communication technologies, small screens are increasingly used for a wider range of purposes. Examples are mobile phones and personal organisers that can be used to compose and edit text, view images or connect to the Internet. Any prolonged use of such devices for work purposes will be subject to the DSE Regulations and the circumstances of such cases will need to be individually assessed. It cannot be assumed that such devices, having much of the functionality of full-sized DSE, are excluded because their screens are 'small'. However, mobile phones that are in prolonged use only for spoken conversation are excluded under regulation 1(4)(e) because their display screens are incidental to this kind of use.

26   The exclusion in regulation 1(4)(f) is for window typewriters having a small display showing no more than a few lines of text.

Application of the DSE Regulations to special groups of workers

Homeworkers and teleworkers

27   If a DSE user is employed to work at home, or at other locations away from their main base, the DSE Regulations apply - whether or not the workstation is provided in whole or in part by the employer. There is no evidence that homeworkers/teleworkers are exposed to any major additional or unique risks to health and safety as a consequence of their DSE work. Indeed, there may be potential benefits to individuals and communities, for example through provision of work to disadvantaged groups and deprived areas, and reduced personal risk and environmental damage from commuting.

28   However, homeworkers/teleworkers may encounter both the normal risks associated with DSE work and some potentially increased risks that may arise from social isolation, stress, lack of supervision, lack of employer commitment and difficulties in undertaking risk assessments. There are some easy steps that should be taken to reduce these risks:

(a)   It is not always practicable for the employer to send someone else to conduct a risk assessment for homeworkers/teleworkers (particularly in the case of mobile workers). A good solution is for the employer to train such workers to undertake their own risk assessments. This can be done by training homeworkers/teleworkers to use an ergonomic checklist, like the one in Appendix 5. Such training could be done before homeworking starts or when the employee concerned is visiting head office, using office DSE to work through the topics an assessment should cover. See also item (d) below which points out there needs to be a clear understanding about who has the responsibility for dealing with any risks found.

(b)   In the case of the homeworker or teleworker's main workstation, the procedure set out in paragraph 28(a) should be used to undertake a full risk assessment which should be reported to the employer, for example by submitting the completed checklist. In the case of mobile teleworkers working for very short periods in hotel rooms and similar places, the full procedure may not be necessary. However, such workers should be trained to mentally run through key points from the checklist, and take appropriate steps to ensure they are comfortable and can minimise the risks wherever and whenever they carry out DSE work.

(c)   In addition to training in risk assessment, homeworkers/teleworkers will need extra training and information about health and safety relating to DSE use (for example good posture, taking breaks). This is important for all users, but is especially so for homeworkers/teleworkers, who are not under immediate supervision and are also unable to pick up good habits by watching colleagues. Break-monitoring software (see paragraphs 65-67) may be worth considering.

(d)   Homeworkers/teleworkers will need clear chains of communication for reporting and resolving any ergonomic defects or warning signs of health problems. This works best if there is a supportive workplace culture so that users feel encouraged to report back promptly about any problems encountered, and users and managers are motivated to find and implement solutions.

29   Where homeworkers/teleworkers are using portable DSE, refer to the guidance in Appendix 3. HSE's guidance *Homeworking*[8] deals with general risk assessments of the home environment and the extent to which the employer is responsible for home electrical systems and equipment.

Agency workers

30   Many temporary workers supplied by employment agencies/employment businesses will use DSE sufficiently to become users (employees) or operators (self-employed people) and hence be subject to the DSE Regulations.

31   Where a DSE worker supplied by an agency/employment business becomes an employee of the host employer, the duties under the DSE Regulations will fall to the host employer. In other situations where the worker is an employee of the agency or is self-employed, both the agency/employment business and the host (client) employer will have duties under the DSE Regulations. The following list clarifies these responsibilities:

(a)   Host employers should:

(i)   assess risks to agency workers (whether users or operators) using their workstations (regulation 2);

11

(ii) ensure all workstations in their undertaking comply with minimum requirements (regulation 3);

(iii) ensure activities are planned so that agency worker users can have breaks from DSE work (regulation 4);

(iv) provide training to agency worker users when their workstation is being modified (regulation 6(2));

(v) provide information to agency workers (both users and operators) about risks, risk assessment and risk reduction measures; and additionally to users about breaks, and training when their workstation is modified (regulation 7).

(b) Employment businesses (agencies) should:

(i) on request, provide eye tests (and special corrective appliances, if required) to agency worker users who are their employees (regulation 5);

(ii) provide health and safety training for such workers (regulation 6(1) and 6(1A));

(iii) provide information to such workers about eye tests and training (regulation 7);

(iv) check that host employers carry out their duties (as described above) to:

● conduct risk assessments of the workstations to be used;

● ensure their workstations comply with the minimum requirements;

● plan for breaks or changes of activity for users; and

● provide information to agency workers, as specified above.

Analysis of workstations to assess and reduce risks

(1)   Every employer shall perform a suitable and sufficient analysis of those workstations which –

(a)   (regardless of who has provided them) are used for the purposes of his undertaking by users; or

(b)   have been provided by him and are used for the purposes of his undertaking by operators,

for the purpose of assessing the health and safety risks to which those persons are exposed in consequence of that use.

(2)   Any assessment made by an employer in pursuance of paragraph (1) shall be reviewed by him if –

(a)   there is reason to suspect that it is no longer valid; or

(b)   there has been a significant change in the matters to which it relates;

and where as a result of any such review changes to an assessment are required, the employer concerned shall make them.

(3)   The employer shall reduce the risks identified in consequence of an assessment to the lowest extent reasonably practicable.

(4)   The reference in paragraph (3) to "an assessment" is a reference to an assessment made by the employer concerned in pursuance of paragraph (1) and changed by him where necessary in pursuance of paragraph (2).

32   Possible health risks which have been associated with DSE work are described in Appendix 2. This appendix also sets the risks in the context of the Priority Programmes to reduce musculoskeletal disorders and stress, and discusses the principles of successful prevention, treatment and rehabilitation.

33   The principal risks relate to physical (musculoskeletal) problems, visual fatigue and mental stress. These are not unique to DSE work nor an inevitable consequence of it, and indeed there is some evidence that the risk to the individual user from typical DSE work is low if appropriate precautions are taken. However, in DSE work as in other types of work, ill health can result from poor equipment or furniture, work organisation, working environment, job design and posture, and from inappropriate working methods.

34   The risk of musculoskeletal disorders in DSE work can be increased by the work-related factors listed above, by home or leisure activities, or by a combination of the two. While surveys indicate that only a small proportion of DSE workers are likely to suffer significant ill health, the number of cases is still far from negligible as DSE workers are so numerous. However, the known health problems associated with DSE work can be prevented in the majority of cases by good ergonomic design of the equipment, workplace and job, and by worker training and consultation.

35   Employers will need to assess the extent to which any of the above risks arise for DSE workers using their workstations who are:

(a)   users employed by them;

(b)   users employed by others (for example agency employed 'temps');

(c)   operators, ie self-employed contractors who would be classified as users if they were employees (for example self-employed agency 'temps', self-employed journalists).

36   Individual workstations used by any of these people will need to be analysed and risks assessed. If employers require their employees to use workstations at home, these too will need to be assessed (see paragraph 28). If there is doubt whether any individual is a user or operator, carrying out a risk assessment may help in reaching a decision.

Suitable and sufficient analysis and risk assessment

37   Risk assessment should first identify any hazards and then evaluate risks and their extent. A **hazard** is something with the potential to cause harm; **risk** expresses the likelihood that the harm from a particular hazard is realised. The **extent of the risk** takes into account the number of people who might be exposed to a risk and the consequences for them. Risks to health may arise from a combination of factors and are particularly likely to occur when the work, workplace and work environment do not take account of workers' needs. Therefore, a suitable and sufficient analysis should:

(a) be systematic, including investigation of non-obvious causes of problems. For example poor posture may be a response to screen reflections or glare, rather than poor furniture;

(b) be appropriate to the likely degree of risk. This will largely depend on the duration, intensity or difficulty of the work undertaken, for example the need for prolonged high concentration because of particular performance requirements;

(c) be comprehensive, considering both:

(i) the results of analysis of the workstation (equipment, furniture, software and environment); and

(ii) organisational and individual factors, including things like workloads and working patterns, provision of breaks, training and information, and any special needs of individuals (such as people with a disability);

(d) incorporate information provided by both employer and worker; and

(e) include a check for the presence of desirable features as well as making sure that bad points have been eliminated.

The form of the assessment

38 In the simplest and most obvious cases which can be easily repeated and explained at any time, an assessment need not be recorded. This might be the case, for example if no significant risks are found and no individual user or operator is identified as being especially at risk. Assessments of short-term or temporary workstations may also not need to be recorded, unless risks are significant. However, in most other cases assessments need to be recorded and kept readily accessible to ensure continuity and accuracy of knowledge among those who may need to know the results (for example where risk reduction measures have yet to be completed). Recorded assessments need not necessarily be a paper and pencil record but can be stored electronically.

39 Information provided by users is an essential part of an assessment. The inclusion of such views is likely to result in better information on existing conditions and provide a feeling of ownership over the findings. Employees who are actively involved in the risk assessment process are also more likely to report any problems as they arise. A useful way of obtaining this information can be through an ergonomic checklist, which should preferably be completed by users or with their input. An example of such a checklist is given in Appendix 5. Other approaches are also possible. For example more objective elements of the analysis (for example chair adjustability, keyboard characteristics, nature of work, etc) could be assessed generically in respect of particular types of equipment - or groups of workers performing the same tasks. Other aspects of workstations would still need to be assessed individually through information collected from users, but this could then be restricted to subjective factors (for example relating to comfort, adjustability of chairs, particularly where there is hot-desking, etc). **Whatever type of checklist is used, employers should ensure workers have received the necessary training before being asked to complete one**.

40 The form of the assessment needs to be appropriate to the nature of the tasks undertaken and the complexity of the workstation. For many office tasks the assessment can be a judgement based on responses to the checklist. Where particular risks are apparent however, and for complex situations, for example

14

where safety of others is a critical factor, a more detailed assessment may be appropriate. This could include, for example a task analysis where particular job stresses had been identified, recording of posture, physical measurement of workstations, or quantitative surveys of lighting and glare.

Shared workstations

41   Where one workstation is used by more than one worker, whether simultaneously or in shifts, it should be analysed and assessed in relation to all those covered by the DSE Regulations. For example if a very tall and a very short worker are sharing a workstation, the assessor should check the chair has a wide enough range of adjustment to accommodate both of them, and that a footrest is available when required.

Who should do assessments?

42   Assessments can be made by health and safety personnel, or other in-house staff, if they have (or are trained for) the abilities required (see paragraph 43). It may be necessary to call in outside expertise where, for example DSE or associated components are faulty in design or use, where workstation design is complex, or where critical tasks are being performed. When in-house personnel are trained to act as assessors, suitable checks should be made afterwards that assessors have understood the information given to them and have reached an adequate level of competence. (One way to do this would be for the trainer to check a sample of the assessor's work.)

43   Those responsible for the assessment should be familiar with the main requirements of the DSE Regulations and should have the ability to:

(a)   identify hazards (including less obvious ones) and assess risks from the workstation and the kind of DSE work being done; for example by completing a checklist or reviewing one completed by the worker;

(b)   draw upon additional sources of information on risk as appropriate;

(c)   draw valid and reliable conclusions from assessments and identify steps to reduce risks;

(d)   make a clear record of the assessment and communicate the findings to those who need to take appropriate action, and to the worker concerned;

(e)   recognise their own limitations as to assessment so that further expertise can be called on if necessary.

44   The inclusion of the views of individual users about their workstations is an essential part of the assessment (as noted in paragraph 39). Employees' **safety representatives** should also be encouraged to play a full part in the assessment process. In particular they should be encouraged to report any problems in DSE work that come to their attention.

Review of assessment

45   The assessment or relevant parts of it should be reviewed in the light of changes to the DSE worker population, or changes in individual capability, or where there has been some significant change to the workstation such as:

(a)   a major change to software used;

(b)   a major change to any of the equipment (screen, keyboard, input devices, etc);

(c) a major change in workstation furniture;

(d) a substantial increase in the amount of time required to be spent using DSE;

(e) a substantial change in other task requirements (for example greater speed or accuracy);

(f) if the workstation is relocated (even if all equipment and furniture stays the same);

(g) if major features of the work environment, such as the lighting, are significantly modified.

46   Reassessments should be done in the same way as the original assessment, consulting workers and safety representatives, and should be done as soon as reasonably practicable after the need for one is identified.

47   Encouraging users to report any ill health that may be due to their DSE work is a useful check that risk assessment and reduction measures are working properly. Reports of ill health may indicate that reassessment is required. Individuals vary in their willingness to report ill health and it is important for employers to explain the benefits of early reporting of any symptoms. The need to report and the organisational arrangements for making a report should be covered in training.

Reducing risks

48   The assessment will highlight any particular areas which may give reason for concern, and these will require further evaluation and corrective action as appropriate. Risks identified in the assessment must be remedied as quickly as possible. For typical applications of DSE, such as computers in offices, remedial action is often straightforward, for example:

(a) **Postural problems** may be overcome by simple adjustments to the workstation such as repositioning equipment or adjusting the chair. Postural problems can also indicate a need to reinforce the user's training (for example on correct hand position, posture, how to adjust equipment). New equipment such as a footrest or document holder may be required in some cases.

(b) **Visual problems** may also be tackled by straightforward means such as repositioning the screen or using blinds to avoid glare, placing the screen at a more comfortable viewing distance from the user, or by ensuring the screen is kept clean. In some cases, new equipment such as window blinds or more appropriate lighting may be needed (see also guidance on regulation 5, on eyes and eyesight).

(c) **Fatigue and stress** may be alleviated by correcting obvious defects in the workstation as indicated in items (a) and (b) above, and ensuring the software is appropriate to the task. In addition, as in other kinds of work, good design of the task will be important. Wherever possible, the task should provide users with a degree of personal control over the pace and nature of their tasks (see guidance on regulation 4). Proper provision must be made for training and information, not only on health and safety risks but also on the use of software. Further advice on software is given in paragraphs 42-44 of Appendix 1.

49   It is important to take a systematic approach to risk reduction and recognise the limitations of the basic assessment. Observed problems may reflect the interaction of several factors or may have causes that are not obvious. For example backache may turn out to have been caused by the worker sitting in an abnormal position in order to minimise the effects of reflections on the screen. If the factors underlying a problem appear to be complex, or if simple remedial measures do not have the desired effect, it will generally be necessary to obtain expert advice on corrective action.

Sources of information and advice

50   The reference section at the back of the book contains a list of relevant HSE guidance documents, for example on lighting and seating, and other publications. Further advice on health problems that may be connected with display screen work may be available from in-house safety or occupational health departments where applicable or, if necessary, from the Employment Medical Advisory Service (see Appendix 6). Expert advice may be obtained from independent specialists in relevant professional disciplines such as ergonomics or lighting design. The reference section includes some useful publications from relevant professional bodies.

Standards

51   Ergonomic specifications for use of DSE are contained in various international, European and British standards. Further information is given in Appendix 1. Compliance with relevant parts of these standards is not a legal requirement. Such compliance will generally not only satisfy but also go beyond the requirements of the DSE Regulations, because the standards aim to enhance performance as well as health and safety.

Requirements for workstations

Every employer shall ensure that any workstation which may be used for the purposes of his undertaking meets the requirements laid down in the Schedule to these Regulations, to the extent specified in paragraph 1 thereof.

Extent to which employers must ensure that workstations meet the requirements laid down in this Schedule

1   An employer shall ensure that a workstation meets the requirements laid down in this Schedule to the extent that –

(a)   those requirements relate to a component which is present in the workstation concerned;

(b)   those requirements have effect with a view to securing the health, safety and welfare of persons at work; and

(c)   the inherent characteristics of a given task make compliance with those requirements appropriate as respects the workstation concerned.

52   Regulation 3 refers to the Schedule to the DSE Regulations which sets out minimum requirements for display screen workstations, covering the equipment, the working environment, and the interface (for example software) between the computer and the user or operator. Figure 1 summarises the main requirements. Appendix 1 contains more information on those parts of the Schedule which may need interpretation.

53   Regulation 3 was amended in 2002 to take account of a judgment in the European Court of Justice (joined cases C-74 and 129/95); the amended regulation is shown above. The judgment ruled that articles 4 and 5 of the Display Screen Equipment Directive (90/270/EEC) impose obligations in respect of all DSE workstations, not just those used by 'workers' as defined in the Directive. Regulation 3 has therefore been altered to remove references to use by users or operators (which are the defined terms in the DSE Regulations equivalent to 'workers' in the Directive). In amending regulation 3, the opportunity has also been taken to simplify it by removing obsolete references to a transitional period for modification of older equipment (see paragraph 58).

54   Employers should therefore modify all DSE workstations they possess that do not already comply. However, modification is only required to the extent described by paragraph 1 of the Schedule, as explained in paragraph 55.

Application of the Schedule

55   By virtue of paragraph 1 of the Schedule, the minimum requirements set out in paragraphs 2-4 of the Schedule apply only in so far as:

(a)   **the components concerned (for example document holder, chair or desk) are present at the workstation**. Where a particular item is mentioned in the Schedule, this should not be interpreted as a requirement that all workstations should have one, unless risk assessment under regulation 2 suggests the item is necessary;

(b)   **they relate to worker health, safety and welfare**. For the purposes of these Regulations, it is only necessary to comply with the detailed requirements in paragraphs 2, 3 and 4 of the Schedule if this would secure the health, safety or welfare of people at work. The requirements in the Schedule do not extend to the efficiency of use of DSE, workstations or software. However, these matters are covered, in addition to worker health and safety, in standards such as BS EN ISO 9241[9] (see Appendix 1). Compliance with standards is not a legal requirement (see paragraph 3 of Appendix 1), but following the

Figure 1 Subjects dealt with in the Schedule

- Adequate lighting

- Adequate contrast, no glare or distracting reflections

- Distracting noise minimised

- Leg room and clearances to allow postural changes

- Window covering if needed to minimise glare

- Software: appropriate to task, adapted to user, providing feedback on system status, no undisclosed monitoring

- Screen: stable image, adjustable, readable, glare/reflection-free

- Keyboard: usable, adjustable, detachable, legible

- Work surface: with space for flexible arrangement of equipment and documents; glare-free

- Chair: stable and adjustable

- Footrest if user needs one

standards, where they are appropriate, should enhance efficiency as well as ensuring that the relevant health and safety requirements of the Schedule are satisfied.

(c) **the inherent requirements or characteristics of the task make compliance appropriate**. It is not a requirement to comply with all the detailed requirements in paragraphs 2, 3 and 4 of the Schedule if doing so would mean the task for which the workstation is used could not be carried out successfully. Some examples are given in paragraph 57. Note that it is the demands of the task, rather than the capabilities of any particular equipment, that are the deciding factor here.

56   In practice, the detailed requirements in paragraphs 2-4 of the Schedule are most likely to be fully applicable in typical office situations, for example where DSE is used for tasks such as data entry, e-mail or word processing. In more specialised applications, compliance with particular requirements in the Schedule may be inappropriate where there would be no benefit to health and safety (or adverse effects on it). Where DSE is used to control machinery, processes or vehicle traffic, it is clearly essential to consider the implications of any design changes for the rest of the workforce and the public, as well as the health and safety of the screen user.

57   The following examples illustrate how these factors can operate in practice. They each include a reference to the relevant part of paragraph 1 of the Schedule:

(a)   Where, as in some control-room applications, a screen is used from a standing position and without reference to documents, a work surface and chair may be unnecessary (Schedule 1(a)).

(b)   Some individuals who suffer from certain back complaints may benefit from a chair with a fixed back rest or a special chair without a back rest (Schedule 1(b)).

(c)   Wheelchair users work from a 'chair' that may not comply with the requirements in paragraph 2(e) of the Schedule. They may have special requirements for work surface (for example height); in practice some wheelchair users may need a purpose-built workstation (or one with height adjustability) but others may prefer to use existing work surfaces. Clearly the needs of the individual here should have priority over rigid compliance with paragraph 2 of the Schedule (Schedule 1(b)).

(d)   Where a user may need to rapidly locate and operate emergency controls, placing them on a detachable keyboard may be inappropriate (Schedule 1(b) and (c)).

(e)   Where there are banks of screens, as in process control or air traffic control, individually tilting and swivelling screens may be undesirable as the screens may need to be aligned with one another and/or be aligned for easy viewing from the operator's seat. Detachable keyboards may also be undesirable if a particular keyboard needs to be associated with a particular screen and/or instrumentation in a multi-screen array (Schedule 1(c) and (b)).

(f)   A brightness control may be inappropriate for process control screens used to display alarm signals: turning down the brightness could cause an alarm to be missed (Schedule 1(b) and (c)).

20

(g) Screens that are necessarily close to other work equipment (for example in a fixed assembly such as a control room panel) that needs to be well-illuminated will need carefully positioned local lighting. It may then be inappropriate for the screen to tilt and swivel as this could give rise to strong reflections on the screen (Schedule 1(b)).

(h) Where microfiche is used to keep records of original documents, screen characters may not be well-defined or clearly formed if the original was in poor condition or was badly photographed (Schedule 1(c)).

(i) Radar screens used in air traffic control may have characters which have blurred 'tails' and hence might be considered to be not well-defined and clearly formed; however, long-persistence phosphors are deliberately used in such screens in order to indicate the direction of movement of the aircraft (Schedule 1(c)).

(j) Screens forming part of a simulator for training the crews of vehicles (ships, trains or aircraft) may have special features that do not comply with the Schedule but are necessary if the simulator is to accurately mimic the features of the (exempt) DSE on the vehicle (Schedule 1(c)).

Transitional period and exclusions

58 Regulation 3 originally contained transitional provisions which allowed employers more time to modify workstations that were already in service before January 1993. However, this transitional period expired on 31 December 1996 and hence is no longer relevant. Employers are now required to ensure that all workstations, whether or not they are new, comply with the Schedule where it is relevant.

59 Where the Schedule does not apply because its requirements are not applicable (under paragraph 1), employers must still comply with other provisions of the DSE Regulations as well as with the HSW Act to ensure that risks to users and operators are reduced to the lowest extent reasonably practicable. So:

(a) if assessment of an existing workstation shows there is a risk to users or operators, the employer should take immediate steps to reduce the risk; or

(b) where paragraph 1(a) or (c) of the Schedule is applicable and the minimum requirements in paragraphs 2, 3 and 4 of the Schedule are therefore not being followed, the employer must ensure that the health and safety of users and operators are adequately safeguarded by whatever other means are appropriate, reasonably practicable and necessary.

Daily work routine of users

Every employer shall so plan the activities of users at work in his undertaking that their daily work on display screen equipment is periodically interrupted by such breaks or changes of activity as reduce their workload at that equipment.

60   In many tasks, natural breaks or pauses occur as a consequence of the inherent organisation of the work. Whenever possible, jobs using DSE should be designed to consist of a mix of screen-based and non-screen-based work to prevent fatigue and to vary visual and mental demands. Where the job unavoidably contains spells of intensive DSE work (whether using the keyboard or input device, reading the screen, or a mixture of the two), these should be broken up by periods of non-intensive, non-DSE work. Where work cannot be so organised, for example in jobs requiring only data or text entry or screen monitoring requiring sustained attention and concentration, deliberate breaks or pauses must be introduced.

Nature and timing of breaks or changes of activity

61   Where the DSE work involves intensive use of the keyboard, mouse or other input device, any activity that would demand broadly similar use of the arms or hands should be avoided during breaks. Similarly, if the DSE work is visually demanding any activities during breaks should be of a different visual nature. Breaks must also allow users to vary their posture. Exercise routines (for example body stretches, blinking the eyes and focusing on distant objects) can be helpful and could be covered in training programmes. Such stretching movements or exercises can help to combat negative effects (such as reduced blood flow) arising from the sedentary nature of most DSE work. Brief stretching exercises can be done whenever necessary, not just in formal breaks.

62   It is not appropriate to lay down requirements for breaks which apply to all types of work; it is the nature and mix of demands made by the job which determine the length of break necessary to prevent fatigue. But some general guidance can be given:

(a)   Breaks or changes of activity should be included in working time. They should reduce the workload at the screen, ie should not result in a higher pace or intensity of work on account of their introduction.

(b)   Breaks should be taken when performance and productivity are still at a maximum, before the user starts getting tired. This is better than taking a break to recover from fatigue. Appropriate timing of the break is more important than its length.

(c)   Short, frequent breaks are more satisfactory than occasional, longer breaks: for example a 5-10 minute break after 50-60 minutes continuous screen and/or keyboard work is likely to be better than a 15-20 minute break every 2 hours.

(d)   Wherever practicable, users should be allowed some discretion as to when to take breaks and how they carry out tasks; individual control over the nature and pace of work allows optimal distribution of effort over the working day.

(e)   Changes of activity (time spent doing other tasks not using the DSE) appear from study evidence to be more effective than formal rest breaks in relieving visual fatigue.

(f)   If possible, breaks should be taken away from the DSE workstation, and allow the user to stand up, move about and/or change posture.

The employer's duty to plan activities

63   The employer's duty under regulation 4 to plan the activities of users can be satisfied by arranging tasks and providing information/training so that users

are able to benefit from breaks or changes of activity; and encouraging them to do so. The duty to plan does not imply a need for the employer to draw up a precise and detailed timetable for periods of DSE work and breaks. That would only be necessary in a few cases.

64   It is generally best for users to be given some discretion over when to take breaks. In such cases the employer's duty to plan activities may be satisfied by allowing an adequate degree of flexibility for the user to organise their own work. However, users given total discretion may forgo breaks in favour of a shorter working day, and thus may suffer fatigue. Employers should ensure that users are given adequate information and training on the need for breaks (see paragraphs 90-100). To gain maximum benefit from breaks, users should be discouraged from using the computer during breaks for any purpose (including their own, for example to surf the Internet). Where users do not take proper breaks despite being trained, it may be necessary for employers to lay down minimum requirements for breaks while still allowing users some flexibility.

65   A number of break-monitoring software tools are marketed as aids to ensure users take regular breaks. They are by no means essential but may be worth considering in some situations. If they are used, the employer still has a responsibility to ensure that work activities are properly planned and that the use of the aid does in fact result in appropriate breaks being taken.

66   Such software packages vary in the facilities they offer. The most basic simply remind the user to take a break at preset intervals, regardless of how much they have used the computer. These may provide a simple solution where computer use is fairly constant throughout the work period. More sophisticated packages monitor the number of keystrokes and/or degree of mouse activity and display a reminder when the user reaches a preset threshold (for example number of keystrokes, or keying rate). These may be especially appropriate where computer use is variable but with some intensive use periods. Care may need to be taken in setting the thresholds for such packages and it is worth bearing in mind that they are unable to detect intensive screen reading which may require a time-related reminder.

67   Employers contemplating purchase of break-monitoring software should be aware that some features can add to the frustration and stress experienced by users and so are undesirable. For example the software should not lock the user out of the DSE without giving adequate warning, to allow the user time to reach a suitable point to take a break. Reminders are generally better than being forced to stop. It is best for the user to have some scope to configure the software, so that it prompts them to take breaks at intervals which take into account their individual needs.

68   The employer's duty is to plan activities so that breaks or changes of activity are taken by users during their normal work. There are a few situations, for example where users working in a control room are handling an unforeseen emergency, where other health and safety considerations may occasionally dictate that normal breaks are not taken.

4

Regulation 5

Eyes and eyesight

(1)   Where a person –

(a)   is a user in the undertaking in which he is employed; or

(b)   is to become a user in an undertaking in which he is, or is to become, employed,

the employer who carries on the undertaking shall, if requested by that person, ensure that an appropriate eye and eyesight test is carried out on him by a competent person within the time specified in paragraph (2).

(2)   The time referred to in paragraph (1) is –

(a)   in the case of a person mentioned in paragraph (1)(a), as soon as practicable after the request; and

(b)   in the case of a person mentioned in paragraph (1)(b), before he becomes a user.

(3)   At regular intervals after an employee has been provided (whether before or after becoming an employee) with an eye and eyesight test in accordance with paragraphs (1) and (2), his employer shall, subject to paragraph (6), ensure that he is provided with a further eye and eyesight test of an appropriate nature, any such test to be carried out by a competent person.

(4)   Where a user experiences visual difficulties which may reasonably be considered to be caused by work on display screen equipment, his employer shall ensure that he is provided at his request with an appropriate eye and eyesight test, any such test to be carried out by a competent person as soon as practicable after being requested as aforesaid.

(5)   Every employer shall ensure that each user employed by him is provided with special corrective appliances appropriate for the work being done by the user concerned where –

(a)   normal corrective appliances cannot be used; and

(b)   the result of any eye and eyesight test which the user has been given in accordance with this regulation shows such provision to be necessary.

(6)   Nothing in paragraph (3) shall require an employer to provide any employee with an eye and eyesight test against that employee's will.

69   In 2002 regulations 5(1) and 5(2) were replaced and regulation 5(3) was amended to improve clarity and remove any doubt over what is required. The amended version is shown above.

70   The purpose of providing eye tests for DSE users is to enhance comfort and efficiency by identifying and correcting vision defects, thus helping to prevent temporary eyestrain and fatigue. There is no reliable evidence that work with DSE causes any **permanent** damage to eyes or eyesight, but it may make users with pre-existing vision defects more aware of them. This (and/or poor working conditions) may give some users temporary visual fatigue or headaches. Uncorrected vision defects can make work at display screens more tiring or stressful than it should be, and correcting defects can improve comfort, job satisfaction and performance. (Note that some DSE work may also require specific visual capabilities such as colour discrimination.)

Eye and eyesight test

71   Regulations 5(1) and 5(2) require employers to provide users who so request it with an appropriate eye and eyesight test. In Great Britain an 'appropriate eye and eyesight test' means a 'sight test' as defined in the

Opticians Act legislation.* The test includes a test of vision and an examination of the eye. For the purpose of the DSE Regulations, the test should take account of the nature of the user's work, including the distance at which the screen is viewed. DSE users are not obliged to have such tests performed. Where users choose to exercise their entitlement, employers should offer an examination by a registered ophthalmic optician, or a registered medical practitioner with suitable qualifications (referred to as optometrist and doctor respectively in the rest of the guidance). (All registered medical practitioners, including those in company occupational health departments, are entitled to carry out sight tests but normally only those with an ophthalmic qualification do so.)

72   Regulation 5(1) gives employers a duty to ensure the provision of appropriate eye and eyesight tests on request:

(a)   to their employees who are already users; and

(b)   to people who are being recruited to be a user (including existing employees being transferred to work that will make them a user).

73   Regulation 5(2) sets out when tests have to be provided to different categories of people:

(a)   Where an existing user requests a test for the first time, the employer should arrange for a test to be carried out as soon as practicable.

(b)   Where the test is requested by an employee who is not yet a user but is to become one, the employer should arrange for a test to be carried out before the person concerned becomes a user.

(c)   Employers must provide eye and eyesight tests on request to any person being recruited as a user. This duty arises only when it is certain that any such person is to become both a user and an employee. Employers do not have to provide eye and eyesight tests to applicants for jobs, though employers may do so if they wish. Where a test is requested by a person being recruited to be a user, if the test is not provided before the person takes up the job it must be provided before the new employee undertakes sufficient DSE work to make them a user. Employers may not refuse to provide a test on the grounds that a new recruit has recently had one provided in any previous period of employment. (However, it may not be of practical benefit to such a user to request a test, if their new tasks and work environment are to be similar to those before the change of job.)

74   The College of Optometrists has produced a statement of good practice for optometrists on DSE matters, published in the guidance for professional conduct on their website (see Appendix 6). Among other things, it makes clear that the purpose of the eye test by an optometrist or doctor under regulation 5 is to decide whether the user has any defect of sight which requires correction when working with a display screen. It follows that when they have an eye test, users need to be able to describe their display screen and working environment (particularly the distance at which they view the screen). As the College points out, the optometrist may need to make a report to the employer, copied to the employee, stating clearly whether or not a corrective appliance is needed

---

*Section 36(2) of the Opticians Act 1989 defines testing sight as 'testing sight with the object of determining whether there is any and, if so, what defect of sight and of correcting, remedying or relieving any such defect of an anatomical or physiological nature by means of an optical appliance prescribed on the basis of the determination'. Further information is given in the Sight Testing Examination and Prescription (No 2) Regulations 1989/1230, which require a doctor or optician to perform specified examinations to detect injury, disease or abnormality when carrying out an eye test.

specifically for display screen work and when re-examination should take place. Any prescription, or other confidential clinical information from the eye test, can only be provided to the employer with the employee's consent.

75   Employers should tell users they employ about the arrangements they have made to provide eye tests to those who want them (there is a requirement to provide this information under regulation 7; see paragraph 96 of the main guidance).

Vision screening tests

76   Vision screening tests are not an 'eye and eyesight test' and hence do not satisfy the DSE Regulations, but some employers may wish to offer them as an extra. Vision screening is a means of identifying individuals with defective vision; however, screening is not designed to find those eye defects, such as injury or disease, that may not at first affect vision. Where companies offer this facility, some users may be content with a vision screening test to check their need for a full sight test. However, employers must also provide the full eye and eyesight test specified in paragraph 71 to those users who either choose at the outset to exercise their entitlement to the full test, or choose to do so after having had vision screening.

77   Where vision screening is offered, the screening instrument or other test method used should be capable of testing vision at the distances appropriate to the user's display screen work, including the intermediate distance at which screens are viewed (normally 50-60 cm). Where test results indicate that vision is defective at the relevant distances, the user should be informed and referred to an optometrist or doctor for a full sight test.

78   Those conducting eyesight screening tests should have basic knowledge of the eye and its function and be competent in operation of the instrument and/or tests. Both the test results and the need for further referral should be assessed by those with medical, ophthalmic, nursing or paramedical skills.

Regularity of provision of eye and eyesight tests

79   Regulation 5 requires that eye and eyesight tests are provided:

(a)   as soon as practicable after display screen users have made a request;

(b)   for recruits or employees who are to become users, and have made a request. In such cases the test must be carried out before the employee becomes a user;

(c)   for users at regular intervals after the first test, provided that they want the tests. Employers should be guided by the clinical judgement of the optometrist or doctor on the frequency of repeat testing. The frequency of repeat testing needed will vary between individuals, according to factors such as age. Employers are not responsible for any corrections for vision defects or examinations for eye complaints which are not related to display screen work which may become necessary within the period. These are the responsibility of the individual concerned;

(d)   for users experiencing visual difficulties which may reasonably be considered to be related to the display screen work, for example visual symptoms such as eyestrain or focusing difficulties.

80   Where an eye test by an optometrist suggests that a user is suffering eye injury or disease, the user will be referred to his or her registered medical practitioner for further examination. This examination is free of charge under the National Health Service.

Corrective appliances

81   'Special' corrective appliances (normally spectacles) provided to meet the requirements of the DSE Regulations will be those appliances prescribed to correct vision defects at the viewing distance or distances used specifically for the display screen work concerned. 'Normal' corrective appliances are spectacles prescribed for any other purpose. It should be noted that experience has shown that in most working populations only a minority (usually less than 10%) will need special corrective appliances for display screen work. Those who need special corrective appliances may include users who already wear spectacles or contact lenses, or others who have uncorrected vision defects.

82   It is good practice to bring other things that need to be viewed for the work into the same visual plane as the screen wherever possible, for example by using a document holder. Hence in many cases where a user requires special glasses to carry out their DSE work, a single lens prescription will be appropriate. However, there may be some circumstances where bifocal or varifocal lenses may seem necessary. These may include situations where the user is required to mix their DSE tasks with other tasks (for example dealing with people) which require a different viewing distance. If the user would repeatedly have to change from one pair of spectacles to another to cope with this, a pair with multiple-focus lenses might be a solution.

83   However, caution is necessary in reaching a decision. There can be side-effects associated with the use of multi-focal prescriptions for DSE work. The smaller size of each lens section can lead to the user having to make repeated adjustments to their neck/head position, or adopting an awkward position in order to look through the appropriate part of the lens. These disadvantages could prove to be more problematic than swapping spectacles, for instance by inducing neck pain. The decision on which is the most suitable solution is best taken in discussion with the optometrist; this will require information being provided about the nature of the work and the workstation and workplace layout.

84   Anti-glare screens, and so-called 'VDU spectacles' and other devices that purport to protect against radiation, are **not** special corrective appliances (see paragraphs 36-39 of Appendix 1 for advice on radiation).

Employers' liability for costs

85   The provision of eye and eyesight tests and of special corrective appliances under the DSE Regulations is at the expense of the **user's employer**. This is the case even if the user works on other employers' workstations.

86   The duty on employers is to **provide** a test where a user requests one. It is up to the employer to decide how to do this, whether by arranging for all their users to visit a particular optometrist or doctor nominated by the employer; by allowing users to make their own arrangements with optometrists and reimbursing the costs afterwards; by a voucher scheme; or any other means.

87   'Normal' corrective appliances are at the user's own expense, but users needing 'special' corrective appliances will be prescribed a special pair of

spectacles for display screen work. Employers' liability for the cost of these is restricted to payment of the cost of a basic appliance, ie of a type and quality adequate for the user's work. Where bifocal or varifocal spectacles are prescribed as special corrective appliances (see caution at paragraph 83) the employer is required to meet the costs associated with providing a basic frame and the prescribed lenses.

88   If, however, users are permitted by their employers to choose spectacles to correct eye or vision defects for purposes which include the user's work but go wider than that, employers need contribute only the costs attributable to the requirements of the job.

89   If users wish to choose more costly appliances (for example with designer frames, or lenses with optional treatments not necessary for the work), the employer is not obliged to pay for these. In these circumstances employers may either provide a basic appliance as above, or may opt to contribute a portion of the total cost of a luxury appliance equal to the cost of a basic appliance.

Provision of training

(1)   Where a person –

(a)   is a user in the undertaking in which he is employed; or

(b)   is to become a user in the undertaking in which he is, or is to become, employed,

the employer who carries on the undertaking shall ensure that he is provided with adequate health and safety training in the use of any workstation upon which he may be required to work.

(1A) In the case of a person mentioned in sub-paragraph (b) of paragraph (1) the training shall be provided before he becomes a user.

(2)   Every employer shall ensure that each user at work in his undertaking is provided with adequate health and safety training whenever the organisation of any workstation in that undertaking upon which he may be required to work is substantially modified.

90   Like regulation 5, regulation 6 was amended in 2002 to remove ambiguities. The amended version is shown above. Employers' duties under it are substantially unchanged, but the regulation now sets out more clearly when training should be provided. Newly recruited users, and existing employees whose duties are changing in a way that will make them become users, should be given training before they start doing the work that will make them a user.

91   Employers should ensure that all users (whether they make use of the employer's workstations or are required to use other workstations) have been provided with adequate and suitable **health and safety** training, in addition to the training received in order to do the work itself. It is good practice for this training to be given before users take part in risk assessments.

92   In practice, there may be considerable overlap between general training requirements and specific health and safety ones (for example the development of keyboard skills) and they are best done together. They will then reinforce each other and facilitate efficient and effective use of the equipment as well as

avoidance of risk. The purpose of training is to increase the user's competence to use workstation equipment safely and reduce the risk to their or anyone else's health. In considering the extent of any training which will be necessary in a particular case, the employer needs to make up any shortfalls between the user's existing competence and that necessary to use the equipment in a safe and healthy way. The development of specific statements of what the user needs to do and how well they need to do it (ie statements of competence) will assist the employer to determine the extent of any shortfall.

93   Training will need to be adapted to the requirements of the particular DSE tasks, be adapted to users' skills and capabilities and be refreshed or updated as the hardware, software, workstation, environment or job are modified. (A workstation should be regarded as having been 'substantially modified' for the purposes of regulation 6(2) if there has been a significant change to it, as set out in paragraph 45.) Where people have been absent from work for long periods, consider if special training or retraining is needed as part of their rehabilitation, particularly if they have suffered from visual, musculoskeletal or stress-related ill health. Organisations should develop systems for identifying the occasions when any of these needs for training arise.

94   Health and safety training should be aimed at reducing or minimising the three risk areas outlined in paragraph 33 of the main guidance and in Appendix 2, with reference to the part played by the individual user. To do this, six interrelated aspects of training should be covered:

(a)   The user's role in correct and timely detection and recognition of hazards and risks. This should cover both the absence of desirable features (for example seat height adjustment) and the presence of undesirable ones (for example screen reflections and glare), together with information on health risks and what to look out for as early warning of problems.

(b)   A simple explanation of the causes of risk and the mechanisms by which harm may be brought about, for example poor posture leading to static loading on the musculoskeletal system and eventual fatigue and discomfort.

(c)   User-initiated actions and procedures which will bring risks under control. Training should cover the following:

(i)   the desirability of comfortable posture and the importance of frequently changing position;

(ii)   correct use of adjustment mechanisms on equipment, particularly furniture, so that stress and fatigue can be minimised;

(iii)   the use and arrangement of workstation components to facilitate good posture, prevent overreaching and avoid glare and reflections on the screen;

(iv)   the need for regular cleaning of screens and other equipment, and inspections to pick up defects requiring maintenance;

(v)   the need to take advantage of breaks and changes of activity.

(d)   Organisational arrangements by which users and their supervisors can alert management to ill health symptoms or problems with workstations.

(e)   Information on these Regulations, particularly as regards eyesight, rest pauses, and the things described in Appendix 1.

29

(f)   The user's contribution to assessments.

95   New users can be given such training at the same time as they are trained on how to use the equipment. The information required to be provided under regulation 7 will reinforce the training and could usefully be in the form of posters or cards with pictorial reminders of some of the essential points. Figure 2 provides an example.

Regulation 7

Provision of information

(1)   Every employer shall ensure that operators and users at work in his undertaking are provided with adequate information about –

(a)   all aspects of health and safety relating to their workstations; and

(b)   such measures taken by him in compliance with his duties under regulations 2 and 3 as relate to them and their work.

(2)   Every employer shall ensure that users at work in his undertaking are provided with adequate information about such measures taken by him in compliance with his duties under regulations 4 and 6(2) as relate to them and their work.

(3)   Every employer shall ensure that users employed by him are provided with adequate information about such measures taken by him in compliance with his duties under regulations 5 and 6(1) as relate to them and their work.

96   Under regulation 7 of the DSE Regulations specific information should be provided as in Table 3.

97   The information should among other things include reminders of the measures taken to reduce the risks such as the system for reporting problems, the availability of adjustable window covering and furniture, **and of how to make use of them**. Providing information will help to consolidate training provided to new users and act as a reminder to those trained previously.

Table 3
Information the employer has to provide to users and operators

	Does employer have to provide information on:					
Category of person working with DSE in the employer's undertaking	*Risks from display screen equipment and workstations?*	*Risk assessment and measures to reduce the risks (regulations 2 and 3)?*	*Breaks and activity changes (regulation 4)?*	*Eye and eyesight tests (regulation 5)?*	*Initial training (regulation 6(1))?*	*Training when workstation modified (regulation 6(2))?*
Users (employees) employed by the employers own undertaking	Yes	Yes	Yes	Yes	Yes	Yes
Users (employees) on site but employed by another employer	Yes	Yes	Yes	No	No	Yes
Operators (self-employed people)	Yes	Yes	No	No	No	No

Figure 2 Seating and posture for typical office tasks

- Seat back adjustable

- Good lumbar support

- Seat height adjustable

- No excess pressure on underside of thighs and backs of knees

- Foot support if needed

- Space for postural change, no obstacles under desk

- Forearms approximately horizontal

- Wrists not excessively bent (up, down or sideways)

- Screen height and angle to allow comfortable head position

- Space in front of keyboard to support hands/wrists during pauses in keying

98   The required information can be provided to staff in any of a number of ways, including putting it in printed form in a memo, wallchart or in health and safety instructions; making it available on a computer disk, electronic bulletin board or other IT-based system, provided that all staff are suitably trained and can access the information; or by verbal briefings. It can be useful to first consult staff to help decide the best method.

99   There is also a general requirement under the Management of Health and Safety at Work Regulations 1999[3] for employers to provide information on risks to health and safety to all their own employees, as well as to employers of other employees on site, to visiting employees, and to the self-employed.

100   Information given to users and operators should also be provided to safety representatives.

Regulation 8

Exemption certificates

(1)   The Secretary of State for Defence may, in the interests of national security, exempt any of the home forces, any visiting force or any headquarters from any of the requirements imposed by these Regulations.

(2)   Any exemption such as is specified in paragraph (1) may be granted subject to conditions and to a limit of time and may be revoked by the Secretary of State for Defence by a further certificate in writing at any time.

(3)   In this regulation –

(a)   "the home forces" has the same meaning as in section 12(1) of the Visiting Forces Act 1952[a];

(b)   "headquarters" has the same meaning as in article 3(2) of the Visiting Forces and International Headquarters (Application of Law) Order 1965[b]; and

(c)   "visiting force" has the same meaning as it does for the purposes of any provision of Part 1 of the Visiting Forces Act 1952.

[a] *1952 c.7.*
[b] *SI 1965/1536, to which there are amendments not relevant to these Regulations.*

Regulation 9

Extension outside Great Britain

These Regulations shall, subject to regulation 1(4), apply to and in relation to the premises and activities outside Great Britain to which sections 1 to 59 and 80 to 82 of the Health and Safety at Work etc Act 1974 apply by virtue of the Health and Safety at Work etc Act 1974 (Application outside Great Britain) Order 1989[a] as they apply within Great Britain.

[a] *SI 1989/840.*

Minimum requirements for workstations

Schedule

(Which sets out the minimum requirements for workstations which are contained in the Annex to Council Directive 90/270/EEC(a) on the minimum safety and health requirements for work with display screen equipment.)

Extent to which employers must ensure that workstations meet the requirements laid down in this Schedule

1   An employer shall ensure that a workstation meets the requirements laid down in this Schedule to the extent that –

   (a)   those requirements relate to a component which is present in the workstation concerned;

   (b)   those requirements have effect with a view to securing the health, safety and welfare of persons at work; and

   (c)   the inherent characteristics of a given task make compliance with those requirements appropriate as respects the workstation concerned.

Equipment

2   (a)   General comment

   The use as such of the equipment must not be a source of risk for operators or users.

   (b)   Display screen

   The characters on the screen shall be well-defined and clearly formed, of adequate size and with adequate spacing between the characters and lines.

   The image on the screen should be stable, with no flickering or other forms of instability.

   The brightness and the contrast between the characters and the background shall be easily adjustable by the operator or user, and also be easily adjustable to ambient conditions.

   The screen must swivel and tilt easily and freely to suit the needs of the operator or user.

   It shall be possible to use a separate base for the screen or an adjustable table.

   The screen shall be free of reflective glare and reflections liable to cause discomfort to the operator or user.

   (c)   Keyboard

   The keyboard shall be tiltable and separate from the screen so as to allow the operator or user to find a comfortable working position avoiding fatigue in the arms or hands.

(a) OJ No L156, 21.6.90, p.14.

The space in front of the keyboard shall be sufficient to provide support for the hands and arms of the operator or user.

The keyboard shall have a matt surface to avoid reflective glare.

The arrangement of the keyboard and the characteristics of the keys shall be such as to facilitate the use of the keyboard.

The symbols on the keys shall be adequately contrasted and legible from the design working position.

(d) Work desk or work surface

The work desk or work surface shall have a sufficiently large, low-reflectance surface and allow a flexible arrangement of the screen, keyboard, documents and related equipment.

The document holder shall be stable and adjustable and shall be positioned so as to minimise the need for uncomfortable head and eye movements.

There shall be adequate space for operators or users to find a comfortable position.

(e) Work chair

The work chair shall be stable and allow the operator or user easy freedom of movement and a comfortable position.

The seat shall be adjustable in height.

The seat back shall be adjustable in both height and tilt.

A footrest shall be made available to any operator or user who wishes one.

Environment

3 (a) Space requirements

The workstation shall be dimensioned and designed so as to provide sufficient space for the operator or user to change position and vary movements.

(b) Lighting

Any room lighting or task lighting provided shall ensure satisfactory lighting conditions and an appropriate contrast between the screen and the background environment, taking into account the type of work and the vision requirements of the operator or user.

Possible disturbing glare and reflections on the screen or other equipment shall be prevented by co-ordinating workplace and workstation layout with the positioning and technical characteristics of the artificial light sources.

(c) Reflections and glare

Workstations shall be so designed that sources of light, such as windows and other openings, transparent or translucid walls, and brightly coloured fixtures or walls cause no direct glare and no distracting reflections on the screen.

Windows shall be fitted with a suitable system of adjustable covering to attenuate the daylight that falls on the workstation.

(d) Noise

Noise emitted by equipment belonging to any workstation shall be taken into account when a workstation is being equipped, with a view in particular to ensuring that attention is not distracted and speech is not disturbed.

(e) Heat

Equipment belonging to any workstation shall not produce excess heat which could cause discomfort to operators or users.

(f) Radiation

All radiation with the exception of the visible part of the electromagnetic spectrum shall be reduced to negligible levels from the point of view of the protection of operators' or users' health and safety.

(g) Humidity

An adequate level of humidity shall be established and maintained.

Interface between computer and operator/user

4   In designing, selecting, commissioning and modifying software, and in designing tasks using display screen equipment, the employer shall take into account the following principles:

(a)   *software must be suitable for the task;*

(b)   *software must be easy to use and, where appropriate, adaptable to the level of knowledge or experience of the operator or user; no quantitative or qualitative checking facility may be used without the knowledge of the operators or users;*

(c)   *systems must provide feedback to operators or users on the performance of those systems;*

(d)   *systems must display information in a format and at a pace which are adapted to operators or users;*

(e)   *the principles of software ergonomics must be applied, in particular to human data processing.*

Appendix 1

Guidance on workstation minimum requirements

1   The Schedule to the DSE Regulations sets out minimum requirements for workstations, applicable mainly to typical office workstations. As explained in the main guidance (paragraph 55), these requirements are applicable only in so far as the components referred to are present at the workstation concerned; the requirements are not precluded by the inherent requirements of the task, and the requirements relate to worker health, safety and welfare. Paragraphs 56 and 57 of the main guidance give examples of situations in which some aspects of these minimum requirements would not apply.

2   The requirements of the Schedule are in most cases self-explanatory but particular points to note are covered below.

General approach: Use of standards

3   Ergonomic requirements for the use of visual display units in office tasks are contained in BS EN ISO 9241.[9] There is no requirement in the DSE Regulations to comply with this or any other standard. Other approaches to meeting the minimum requirements in the DSE Regulations are possible. These other approaches may be appropriate if special requirements of the task or needs of the user mean that equipment, software, etc that complies with the standards is not suitable. However, employers may find standards helpful as workstations and software satisfying BS EN ISO 9241 would meet and in most cases go beyond the minimum requirements in the Schedule to the DSE Regulations.

4   BS EN ISO 9241[9] covers the ergonomics of design and use of visual display terminals in offices; it is concerned with the efficient use of the equipment as well as with user health, safety and comfort. While drafted in connection with office tasks, many of the general ergonomic recommendations in BS EN ISO 9241 will be relevant to some non-office situations.

5   BS EN ISO 9241[9] is an international standard replacing the earlier, interim British Standard BS 7179 which has now been withdrawn. BS EN ISO 9241 is a multipart standard covering the following:

Part 1   General introduction

Part 2   Guidance on task requirements

Part 3   Visual display requirements

Part 4   Keyboard requirements

Part 5   Workstation layout and postural requirements

Part 6   Guidance on the work environment

Part 7   Requirements for displays with reflections

Part 8   Requirements for displayed colours

Part 9   Requirements for non-keyboard input devices

Part 10   Dialogue principles

Part 11   Guidance on usability

Part 12	Presentation of information
Part 13	User guidance
Part 14	Menu dialogues
Part 15	Command dialogues
Part 16	Direct manipulation dialogues
Part 17	Form filling dialogues

6   While this standard is not formally linked to the Display Screen Equipment Directive, one of its aims is to establish appropriate levels of user health and safety and comfort. Technical data in the various parts of the standard may therefore help employers to meet the requirements laid down in the Schedule to the DSE Regulations. Although some parts of BS EN ISO 9241 (such as Parts 3, 4 and 9) are aimed at manufacturers, they still contain information which may help employers - particularly when deciding what to look for in buying new equipment.

7   There are other standards that deal with requirements for furniture. These include BS 3044,[10] which is a guide to ergonomic principles in the design and selection of office furniture generally. BS EN 527[11] also provides information on the dimensions of work tables and desks suitable for office use and BS EN 1335 Part 1[12] has dimensional standards for chairs. Further information about the relevant British, European and international standards can be obtained from the British Standards Institution (see 'References and further reading').

8   Other more detailed and stringent standards are relevant to certain specialised applications of display screens, especially those where the health or safety of people other than the screen user may be affected. Some examples in particular subject areas are:

(a)   **Process control**
A large number of British and international standards are or will be relevant to the design of display screen interfaces for use in process control - such as the part-published BS EN ISO 11064[13] on the general ergonomic design of control rooms.

(b)   **Applications with machinery safety implications**
BS EN 614 - Ergonomic design principles in safety of machinery.[14]
BS EN ISO 13407 - Human-centred design processes for interactive systems.[15]

(c)   **Safety of programmable electronic systems**
BS EN 61508 - Functional safety of electrical/electronic/programmable electronic safety-related systems.[16]

9   Applications such as these are outside the scope of these guidance notes. Anyone involved in the design of such display screen interfaces, or others where there may be safety considerations for non-users, should seek appropriate specialist advice.

Equipment

Display screen

10   Choice of display screen should be considered in relation to other elements of the work system, such as the type and amount of information required for the task, and environmental factors. Satisfactory results can usually be achieved by making appropriate adjustments, for example those described in paragraphs 14, 15 and 31 of this appendix, to adapt a standard display screen to suit particular requirements or changing environmental conditions. However, in some cases, for example control rooms, it may be necessary to use screens that are custom-designed for a specific task or environment.

11   Sizes of display screens are not specified in the DSE Regulations because both the visual demands of tasks and the requirements of particular users vary a great deal. The screen and the characters or images on it need to be large enough for the user to do their work comfortably. While larger screens can give a better image and display more information, it is worth bearing in mind that very large screens take up more desk space which can be detrimental to comfort or require a larger desk. If there is a requirement for a large screen, choosing a flat-panel screen rather than a traditional cathode-ray tube (CRT) display may help resolve this difficulty.

Display stability

12   Individual perceptions of screen flicker vary and with CRT screen technology it is not technically feasible to eliminate flicker for all users. International standards (such as BS EN ISO 9241, Part 3)[9] specify that screens should appear flicker-free to 90% of users; screens that do this can be regarded as satisfying the minimum requirement in the Schedule. A change to a different display can resolve individual problems with flicker. Electrical or magnetic interference from equipment (for example loudspeakers) too close to the screen can sometimes degrade the screen image. Persistent display instabilities - flicker, jump, jitter or swim - may indicate basic design problems, or defects, and assistance should be sought from suppliers. Flat-panel screens such as liquid crystal displays (LCDs), unlike CRTs, are not usually subject to flicker unless they have not been set up properly during manufacture or installation.

Brightness and contrast

13   Either negative or positive image polarity is acceptable, and each has different advantages as shown in Table 4. The balance of these advantages means that positive polarity is the better choice for most work situations.

14   It is important for the brightness and contrast of the display to be appropriate for ambient lighting conditions; trade-offs between character brightness and sharpness may be needed to achieve an acceptable balance. In many kinds of equipment this is achieved by providing a control or controls which allow the user to make adjustments. Controls may need to be readjusted as ambient lighting changes through the day.

Screen adjustability

15   Adjustment mechanisms allow the screen to be tilted or swivelled to avoid glare and reflections and enable the worker to maintain a natural and relaxed posture. They may be built into the screen, form part of the workstation furniture or be provided by separate screen support devices; they

Table 4 Advantages of negative and positive image polarity

Negative polarity (light characters on a dark background)	Positive polarity (dark characters on a light background)
Flicker less perceptible (important with older cathode-ray tube (CRT) screens)	Reflections less perceptible
Legibility better for those with low-acuity vision	Edges appear sharper
Characters may be perceived as larger than they are	Good workplace lighting is easier to achieve

should be simple and easy to operate. Screen height adjustment devices are not essential but may be a useful means of adjusting the screen to the correct height for the worker, especially if a number of different people use the workstation. Alternatively, screens that are too low may be raised using solid blocks or any suitably rigid support that achieves a comfortable height. The reference in the Schedule to adjustable tables does not mean these necessarily have to be provided.

Glare and reflections

16   Screens are generally manufactured without highly reflective surface finishes and often have built-in anti-glare or non-reflective coatings. However, in adverse lighting conditions, reflection and glare may still be a problem. Advice on this is included under lighting (see paragraphs 28-33 of this appendix).

Cleaning

17   Keeping the screen (and keyboard) clean will aid its legibility, increase user comfort and help meet the requirements of the Schedule.

Keyboard and other input devices

18   Keyboard design should allow workers to locate and activate keys quickly, accurately and without discomfort. The choice of keyboard will be dictated by the nature of the task and determined in relation to other elements of the work system. Hand support may be incorporated into the keyboard for support while keying or at rest, depending on what the worker finds comfortable. Support can also be gained by leaving an adequate space between the keyboard and the front edge of the desk; or may be provided by a separate hand/wrist support on or attached to the work surface.

19   It is not a requirement of the DSE Regulations to provide split or otherwise 'ergonomic' keyboards for all users. However, there may be cases where one of these special designs of keyboard may be worth considering, for example to rehabilitate a worker suffering from upper limb pain.

20   DSE is increasingly being used with other, non-keyboard input devices such as the mouse or trackball. These are covered by the DSE Regulations but only in a general way: the requirement is that such equipment should not be a source of risk. Practical advice on choosing and working with such devices is given in Appendix 4.

Work desk or work surface

21   Work surface dimensions may need to be larger than for conventional non-screen office work, to take adequate account of:

(a)   the range of tasks performed (for example screen viewing, keyboard input, use of other input devices, writing on paper, use of telephone, etc);

(b)   position and use of hands for each task;

(c)   use and storage of working materials and equipment (documents, telephones, etc).

22   Document holders can be useful for work with hard copy, particularly for workers who have to repeatedly look from the screen to a document and back, and for anyone who finds difficulty in refocusing. Where a holder is used by touch-typists it should: position working documents at a height, visual plane and viewing distance similar to those of the screen; be of low reflectance; be stable; and not reduce the readability of source documents. People who have to look at the keyboard to type may find it better to place documents close to the keyboard, for example between the keyboard and the screen.

Work chair

23   The primary requirement here is that the work chair should allow the user to achieve a comfortable position. Seat height adjustments should accommodate the needs of users for the tasks performed. The Schedule requires the seat to be adjustable in height (ie relative to the ground) and the seat back to be adjustable in height (also relative to the ground) and tilt. Provided the chair design meets these requirements and allows the user to achieve a comfortable posture, it is not necessary for the height or tilt of the seat back to be adjustable independently of the seat. Automatic backrest adjustments are acceptable if they provide adequate back support. Chairs with arms are liked by some users; but check the arms do not interfere with freedom of movement, for example by stopping the user getting the chair under the work surface to sit comfortably at the keyboard. Remember users may need training on how to adjust chairs.

24   General health and safety advice and specifications for seating are given in the HSE book *Seating at work*.[17]

25   Footrests may be necessary where individual workers are unable to rest their feet flat on the floor (for example where work surfaces cannot be adjusted to the right height in relation to other components of the workstation). Footrests should not be used when they are not necessary as this can result in poor posture.

Environment

26   Note that the Workplace (Health, Safety and Welfare) Regulations 1992[4] contain minimum environmental requirements for all workplaces, covering space, lighting, heating and ventilation, and HSE has published guidance books on many of these subjects (see 'References and further reading'). Some special considerations for DSE work are discussed in paragraphs 27-39 of this appendix.

Space requirements

27   Prolonged sitting in a static position can be harmful. It is most important that support surfaces for display screen and other equipment and materials used at the workstation should allow adequate clearance for postural changes. This means adequate clearances for thighs, knees, lower legs and feet under the work surface and between furniture components. There should be sufficient space for the worker to sit down and get up without difficulty. The height of the work surface should allow a comfortable position (see Figure 2 on page 31) for the arms and wrists, if a keyboard, mouse or other input device is used.

Lighting, reflections and glare

28   Lighting should be appropriate for all the tasks performed at the workstation, for example reading from the screen, keyboard work, reading printed text, writing on paper. General lighting - by artificial or natural light, or a combination - should illuminate the entire room to an adequate standard. Any supplementary individual lighting provided to cater for personal needs or a particular task should not adversely affect visual conditions at nearby workstations.

Illuminance

29   High illuminances render screen characters less easy to see but improve the ease of reading documents. Where a high illuminance environment is preferred for reading documents or for other reasons, the use of positive polarity screens (dark characters on a light background) has advantages as these can be used comfortably at higher illuminances than can negative polarity screens.

Reflections and glare

30   Problems which can lead to visual fatigue and stress can arise, for example from unshielded bright lights or bright areas in the worker's field of view: from an imbalance between brightly and dimly lit parts of the environment; and from reflections on the screen or other parts of the workstation.

31   Measures to minimise these problems include: shielding, replacing or repositioning sources of light; rearranging or moving work surfaces, documents or all or parts of workstations; modifying the colour or reflectance of walls, ceilings, furnishings, etc near the workstation; altering the intensity of vertical to horizontal illuminance; or a combination of these. It is often easiest to avoid reflections and glare if neither the screen nor the user is directly facing windows or bright lights. Adjust curtains or blinds to prevent unwanted light.

32   Anti-glare screen filters should only be considered as a last resort if other measures fail to solve the problem. They are not likely to improve matters for some modern screens which already have an anti-glare finish. Adding a screen filter can have drawbacks; for example its surfaces may get dirty and make it harder to see the screen.

33   General guidance on minimum lighting standards necessary to ensure health and safety of workplaces is available in the HSE guidance book *Lighting at work*.[18] This does not cover ways of using lighting to maximise task

performance or enhance the appearance of the workplace, although it does contain a bibliography listing relevant publications in this area. Specific and detailed guidance is given in the CIBSE Lighting Guide 3 *The visual environment for display screen use.*[19]

Noise

34   Noise from equipment such as printers at display screen workstations should be kept to levels which do not impair concentration or prevent normal conversation (unless the noise is designed to attract attention, for example to warn of a malfunction). Noise from equipment is best reduced at source by specifying quieter alternatives when ordering replacements. Where this is not practicable or will take time, noise can be reduced by soundproofing or repositioning of the equipment. Sound-insulating partitions between noisy equipment and the rest of the workstation are an alternative. Partitions can also help to reduce the distracting effect of noise from other workers.

Heat and humidity

35   Electronic equipment can be a source of dry heat which can modify the thermal environment at the workstation. Ventilation and humidity should be maintained at levels which prevent discomfort and problems of sore eyes. HSE has produced guidance on thermal comfort, *Thermal comfort in the workplace: Guidance for employers.*[20]

Radiation

36   The Schedule requires radiation, with the exception of the visible part of the electromagnetic spectrum (ie visible light), to be reduced to negligible levels from the point of view of the protection of users' health and safety. In fact, so little radiation is emitted from conventional cathode-ray tube (CRT) designs of DSE that no special action is necessary to meet this requirement (see also paragraphs 21-22 of Appendix 2). LCD flat-panel screens do not emit any electromagnetic radiation, except visible light.

37   Taking CRT displays as an example ionising radiation is emitted only in exceedingly small quantities, so small as to be generally much less than the natural background level to which everyone is exposed. Emissions of ultraviolet, visible and infrared radiation are also very small, and workers will receive much less than the maximum exposures generally recommended by national and international advisory bodies.

38   For radio frequencies, the exposures will also be well below the maximum values generally recommended by national and international advisory bodies for health protection purposes. The levels of electric and magnetic fields are similar to those from common domestic electrical devices. Although much research has been carried out on possible health effects from exposure to electromagnetic radiation, no adverse health effects have been shown to result from the emissions from display screen equipment.

39   Thus it is not necessary, from the standpoint of limiting risk to human health, for employers or workers to take any action to reduce radiation levels or to attempt to measure emissions; in fact the latter is not recommended, as meaningful interpretation of the data is very difficult. There is no need for users to be given protective devices such as anti-radiation screens.

Task design and software

Principles of task design

40   Inappropriate task design can be among the causes of stress at work. Stress jeopardises employee motivation, effectiveness and efficiency and in some cases can lead to significant health problems. The DSE Regulations are only applicable where health and safety rather than productivity is being put at risk. However, employers may find it useful to consider both aspects together, as task design changes put into effect for productivity reasons may also benefit health, and vice versa.

41   In DSE work, good design of the task can be as important as the correct choice of equipment, furniture and working environment. It is advantageous to:

(a)   whenever possible, design jobs in a way that offers users variety, opportunities to exercise discretion, opportunities for learning, and appropriate feedback, in preference to simple repetitive tasks (for example the work of a typist can be made less repetitive and stressful if an element of administrative work is added);

(b)   match staffing levels to volumes of work, so that individual users are not subject to stress through being either overworked or underworked;

(c)   allow users to participate in the planning, design and implementation of work tasks whenever possible.

Principles of software ergonomics

42   In most DSE work the software controls both the presentation of information on the screen and the ways in which the worker can manipulate the information. Thus software design is an important element of task design. For many tasks, off-the-shelf software packages are available and such packages may be entirely adequate provided that users are given sufficient training. However, software that is badly designed or inappropriate for the task will impede the efficient completion of the work and in some cases may cause sufficient stress to affect the health of a user. Involving a sample of users in the purchase or design of software can help to avoid problems, particularly if they have the opportunity to try out alternative solutions side by side.

43   Requirements of the organisation, the task, and the DSE workers concerned should first be established as they provide the basis for designing, selecting and modifying software. If additional information is needed it may be helpful to refer to the standard BS EN ISO 9241[9] which contains information on desirable features of text sizes, colours and software.

44   In many (though not all) applications the main points to consider are:

(a)   *Suitability for the task*

(i)   Software should enable workers to complete the task efficiently, without presenting unnecessary problems or obstacles.

(b)   *Ease of use and adaptability*

(i)   Workers should be able to feel that they can master the system and use it effectively following appropriate training. To achieve this, the

43

software should present a consistent interface (for example the way a particular function is accessed and executed should be similar wherever it occurs), and should be reasonably transparent in its operating methods.

(ii) The dialogue between the system and the worker should be appropriate for the worker's ability.

(iii) Where appropriate, software should enable workers to adapt the user interface to suit their ability level and preferences, for example by adjusting mouse speed and sensitivity.

(iv) The software should protect workers from the consequences of errors, for example by providing appropriate warnings and information and by enabling 'lost' data to be recovered wherever practicable.

(c) *Feedback on system performance*

(i) The system should provide appropriate feedback, which may include error messages; suitable assistance ('help') to workers on request; and clear messages about changes in the system such as malfunctions or overloading.

(ii) Feedback messages should be presented at the right time and in an appropriate style and format. They should not contain unnecessary information.

(d) *Format and pace*

(i) Speed of response to commands and instructions should be appropriate to the task and to workers' abilities.

(ii) Characters, cursor movements and position changes should, where possible, be shown on the screen as soon as they are input.

(e) *Performance monitoring facilities*

(i) Quantitative or qualitative checking facilities built into the software can lead to stress if they have adverse results such as an overemphasis on output speed.

(ii) It is possible to design monitoring systems that avoid these drawbacks and provide information that is helpful to workers as well as managers. However, in all cases workers should be kept informed about the introduction and operation of such systems.

Health effects of DSE work and principles of successful prevention, treatment and rehabilitation

Introduction: Why take action?

1   The principal health risks associated with DSE work are physical (musculoskeletal) problems, visual fatigue and mental stress. These problems often reflect bodily fatigue. None of them are unique to DSE work, nor are they an inevitable consequence of it. Risks to typical users should be low if the DSE Regulations are complied with and ergonomic principles are taken into account in the design, selection, installation and use of the equipment; the design of the workplace; and the organisation of the task.

2   However, it is important not to be complacent about this. DSE workers are so numerous that low risk to the average individual may still equate to many thousands of cases of ill health in the working population. In addition, risks to individuals in a particular workplace may not remain low if control measures are poorly designed from the start, or if circumstances change.

3   It is particularly important to consider the possibility of musculoskeletal disorders or stress arising in DSE work. Both of these problems have been targeted by the Health and Safety Commission (HSC) which has set up Priority Programmes on musculoskeletal disorders and stress in its plans of work (details can be found on www.hse.gov.uk). HSC considers that tackling these conditions is vital to delivering the targets set out in the occupational health strategy, *Securing Health Together* and the Government's *Revitalising Health and Safety* initiative. Taking action is also likely to be cost-effective for employers, for example by reducing sickness absences, improving work performance and avoiding compensation claims.

4   The HSC Priority Programmes on musculoskeletal disorders and stress contain targets for ill health reduction and details of planned activities under the five headings of compliance, continuous improvement, knowledge, skills, and support. Delivery of the targets will depend crucially on the commitment of stakeholders in the health and safety system. Stakeholders include employers, workers, safety representatives, Government agencies, local authorities, employers' associations and trade unions, professional bodies, voluntary organisations and many others. They need to work together to manage musculoskeletal and psychosocial risks.

5   As so many people work with DSE, there is considerable potential for reducing the total burden of occupational ill health by managing DSE risks. Stakeholders can help achieve this by:

(a)   reducing risks of ill health, by complying with the DSE Regulations and guidance;

(b)   encouraging early reporting of symptoms;

(c)   ensuring cases of ill health are managed effectively; and

(d)   reviewing risk assessments as and when necessary.

The main hazards of DSE work

Musculoskeletal disorders

6   Musculoskeletal disorders are the most common form of occupational ill health, estimated in 1995 to be affecting over a million people a year in Great Britain and costing society over £5 billion.

7   A range of musculoskeletal disorders of the arm, hand, shoulder and neck linked to work activities are now described as 'upper limb disorders' (ULDs) or 'work-related upper limb disorders' (WRULDs). These range from temporary fatigue or soreness in the limb to chronic soft tissue disorders such as peritendinitis or carpal tunnel syndrome. Some keyboard operators have suffered occupational cramp. Media reports often refer to some, or all, ULDs as 'repetitive strain injury' (RSI) but this term is not a medical diagnosis and can be confusing. For doctors, recommendations on surveillance case definitions (diagnostic criteria) for ULDs were published in 1998.[20]

8   As with other sedentary tasks, DSE work can also give rise to back pain or make existing back pain worse, particularly if seating is poor or badly adjusted, the workstation has insufficient space or is badly designed, or if workers sit too long without changes of posture and breaks from DSE work. HSE has published advice entitled *Back in work: Managing back pain in the workplace*,[22] and BackCare has also published similar advice (see Appendix 6).

9   The contribution of particular risk factors (for example keying rates) to the onset of any disorder may not be clear. Often a combination of factors is involved, which can include non-work factors such as sports, hobbies or earlier injuries, as well as factors relating to the workstation, task or work environment. Prolonged static posture of the back, neck and head is known to contribute to musculoskeletal problems. Awkward positioning of the hands and wrist (for example unnecessarily bent as a result of poor working technique or inappropriate work height) are further risk factors.

10   Outbreaks of ULDs among keyboard workers have often been associated with high workloads combined with tight deadlines. This helps to illustrate the importance of psychosocial factors in DSE work. There is convincing evidence that workers' psychological response to work and workplace conditions may have as important an influence as physical risk factors on their health in general, and musculoskeletal health in particular. Psychosocial risk factors include the design, organisation and management of work and the overall social environment (the *context* of work) and also the specific elements of the job (the *content* of work). It is very likely that physical and psychosocial risk factors combine, and the greatest benefit will be achieved when both are identified and controlled.

11   Many effects of psychosocial factors may be linked to musculoskeletal disorders via stress-related processes which include direct biochemical and physiological changes in the body. Stress is also a hazard in its own right (see paragraphs 13-16 of this appendix) as individuals may try to cope with stressful demands by behaving in ways that in the long term may be detrimental to health. For example DSE workers are frequently tempted to cope with high workloads or deadlines by shortening or not taking breaks.

12   Thus a wide variety of physical and psychosocial factors contribute to the risk of musculoskeletal disorders in DSE work, and differences between individuals also need to be taken into account (for biological reasons there may be some people who are more or less likely than average to develop a

health problem). All this requires a risk reduction strategy which embraces proper equipment, furniture, training, job design and work planning. Complying with the DSE Regulations, as described in this book, will help achieve this. More detailed advice on the principles of tackling ULDs and helping recovery is given in paragraphs 25-31 of this appendix.

Fatigue and stress

13 Stress is the second most common cause of occupational ill health. Prolonged or particularly intense periods of stress can lead to physical and/or mental illness as well as behavioural changes which can damage health, such as smoking or drinking.

14 Many symptoms described by DSE workers reflect stresses arising from their work. Symptoms may be linked to upper limb or visual problems but there is evidence that stress often contributes as well. In considering the design of jobs to reduce psychosocial risks in DSE work, the following factors are undesirable and should be tackled if possible:

(a)   workers having little control over their work and working methods (including shift patterns);

(b)   tasks requiring high attention and concentration in conditions where the worker has little control over their allocation of effort;

(c)   workers being unable to make full use of their skills;

(d)   workers not being involved in making decisions that affect them;

(e)   being expected to carry out repetitive, monotonous tasks all the time;

(f)   work being system-paced (especially if work rates are being monitored inappropriately);

(g)   demands of the work being perceived as excessive;

(h)   payment systems that encourage working too quickly or with insufficient breaks;

(i)   opportunities for social interaction being limited by work systems;

(j)   high levels of effort not being balanced by sufficient reward (pay, resources, self-esteem, status).

15 All these have been linked with stress in DSE work, although clearly they are not unique to it; however, attributing individual symptoms to particular aspects of a job or workplace can be difficult.

16 Training for managers is important in tackling psychosocial risks; their management style and the way they react to problems or complaints can be influential. Also (as with physical risk factors) psychosocial issues are best addressed with full consultation and involvement of the workforce. The risks to DSE workers can be minimised by following the principles underlying the DSE Regulations and guidance, ie by careful design, selection and arrangement of display screen equipment; good design of the user's workplace, environment and task; and training, consultation and involvement of the user. If more advice on work-related stress is required, HSE has published guidance entitled *Work-related stress: A short guide*[23] and *Tackling work related stress:*

Eye and eyesight effects

17   Medical evidence shows that **using DSE is *not* associated with permanent damage to eyes or eyesight**; nor does it make existing defects worse. However, some workers may experience **temporary** visual fatigue, leading to a range of symptoms such as impaired visual performance (for example blurred vision), red or sore eyes and headaches, or the adoption of awkward posture which can cause further bodily discomfort. Visual symptoms may be caused by:

(a)   staying in the same position and concentrating for a long time;

(b)   poor positioning of the DSE;

(c)   poor legibility of the screen, keyboard or source documents;

(d)   poor lighting, including glare and reflections;

(e)   a drifting, flickering or jittering image on the screen.

18   As with other visually demanding tasks, DSE work does not cause eye damage but it may make workers with pre-existing vision defects more aware of them. Such uncorrected defects can make work with DSE more tiring or stressful than would otherwise be the case, which in turn may lead to increased risk of injury from musculoskeletal disorders.

Other minor or alleged health effects

Epilepsy

19   Work with DSE has not been known to induce epileptic seizures. People suffering from the very rare (1 in 10 000 population) photosensitive epilepsy who react adversely to flickering lights and patterns find they can safely do normal office tasks using a display screen. People with epilepsy who are concerned about DSE work can seek further advice from the Employment Medical Advisory Service at HSE regional offices - see the telephone directory. Epilepsy Action (the new name for the British Epilepsy Association; see Appendix 6) can also offer advice on photosensitive epilepsy and DSE work.

Facial dermatitis

20   Some DSE users have reported facial skin complaints such as occasional itching or reddened skin on the face and/or neck. These complaints are rare and the limited evidence available suggests they may be associated with environmental factors, such as low relative humidity or static electricity, and individual susceptibility.

Electromagnetic radiation

21   Anxiety about radiation emissions from DSE and possible effects on pregnant women was once widespread. However, there is substantial evidence that these concerns are unfounded. The Health and Safety Executive has consulted the National Radiological Protection Board (NRPB), which has the statutory function of providing information and advice on all radiation matters to Government departments, and the advice given in paragraphs 22-24 of this appendix summarises scientific understanding. (Anyone requiring more detail,

should consult the NRPB report *Health effects related to the use of visual display units.*[25])

22   The levels of ionising and non-ionising electromagnetic radiation which are likely to be generated by DSE are well below those set out in international recommendations for limiting risk to human health created by such emissions and the NRPB does not consider such levels to pose a significant risk to health. No special protective measures are therefore needed to protect the health of people from this radiation.

Effects on pregnant women

23   In the 1980s there was public concern about reports of higher levels of miscarriage and birth defects among some groups of DSE workers, allegedly due to electromagnetic radiation. Many scientific studies have been carried out, but taken as a whole their results do not show any link between miscarriages or birth defects and working with DSE. Research and reviews of the scientific evidence will continue to be undertaken.

24   In the light of the scientific evidence, pregnant women do not need to stop work with DSE. However, to avoid stress and anxiety, women who are pregnant or planning children and worried about working with DSE should be given the opportunity to read this guidance. If anyone subsequently is still concerned, they should be given the opportunity to discuss the issues with someone adequately informed of current authoritative scientific information and advice. HSE has published a general guidance book about pregnancy *New and expectant mothers at work: A guide for employers.*[26]

Principles of a successful prevention strategy

25   Compliance with the DSE Regulations by following the advice in this book will help to prevent ill health in DSE workers in the great majority of situations. Many employers will not need to do more than this. However, it is good practice to continue to monitor levels of sickness absence and reported discomfort, as a check that the action taken to reduce risks is continuing to be successful.

26   There may be some instances where ill health still occurs, indicating that further analysis and a more thorough approach is needed to tackle the problem. This may be most likely in the case of work-related ULDs.

27   Comprehensive advice on preventing ULDs is contained in HSE guidance book *Upper limb disorders in the workplace.*[27] This is wide-ranging general guidance that is applicable to many industrial and commercial jobs where ULD risks can be high. However, the principles it describes are also relevant to DSE work. It shows how a structured approach may help compliance with the DSE Regulations in unusual situations with complex challenges. This is specifically illustrated in one of the book's case studies which examines computer use in a news media organisation which had encountered a sudden surge of ULD cases after the introduction of a new system.

28   *Upper limb disorders in the workplace*[27] advocates a seven-stage approach to minimising the risk of ULDs:

(a)   understand the issues and commit to action;

(b)   create the right organisational environment;

(c)   assess the risk of ULDs in your workplace;

(d)   reduce the risk of ULDs;

(e)   educate and inform your workforce;

(f)   manage any episodes of ULDs;

(g)   carry out regular checks on programme effectiveness.

Each of these stages is described in detail and the book[27] also contains case studies, a risk filter and risk assessment worksheet, and medical and legal appendices.

29   The approach described in *Upper limb disorders in the workplace*[27] is compatible in principle with the requirements of the DSE Regulations and in some respects may go beyond them. However, there are some specific requirements in the DSE Regulations that cannot be covered in general guidance on ULDs. That is why it is advisable for most employers with DSE users or operators to start by following the specific guidance in this present book on the DSE Regulations, only going on to work through the management principles set out in *Upper limb disorders in the workplace*[27] if the ordinary approach to complying with the DSE Regulations does not solve all the problems.

Treatment and rehabilitation

30   Adequate control of risk factors by compliance with the DSE Regulations will go a long way to prevent the occurrence of ill health in DSE work, but due to individual differences in the body's response it is not possible to ensure that every possible episode will be prevented. It is therefore recommended that employers should have a system in place to detect and manage any cases of work-related ill health that may arise. Such systems should:

(a)   encourage users to report any symptoms early. Individuals' willingness to do this varies, so it is important to establish a supportive climate in the workplace that emphasises the benefits of early detection of possible harm, to reassure those who report musculoskeletal aches and pains that early detection and treatment will normally avoid any serious problems developing;

(b)   provide appropriate advice for users who report symptoms. Depending on circumstances, this might include reassurance that pain does not necessarily mean harm, advice on risk factors, and/or reviewing the individual's work tasks with them;

(c)   provide for referral to health professionals to obtain appropriate diagnosis, treatment or further advice if symptoms are getting worse or there is other cause for concern;

(d)   help sufferers to continue working, or to return to work after periods of absence or treatment. Things to consider might include altering the job or the workstation, easing the person back into the job by a period of work at a reduced pace, providing alternative work, and providing advice and support.

31   With disorders such as ULDs that may be work-related, the occurrence of a confirmed case should be taken as a prompt to review whether risk assessment and risk reduction measures are adequate. This is especially important if there has been more than one case.

Appendix 3

Work with portable DSE

1   Portable DSE, such as laptop and notebook computers, is subject to the DSE Regulations if it is in prolonged use. This appendix gives practical guidance.

2   Increasing numbers of people are using portable DSE as part of their work. While research suggests that some aspects of using portable DSE are no worse than using full-sized equipment, that is not true of every aspect. The design of portable DSE can include features (such as smaller keyboards or a lack of keyboard/screen separation) which may make it more difficult to achieve a comfortable working posture. Portable DSE is also used in a wider range of environments, some of which may be poorly suited to DSE work.

3   To reduce risks to portable DSE users, the following recommendations should be followed (in addition to following the general advice for all DSE work in the main part of this book).

Risk assessment (regulation 2)

4   Risk assessment for users of portables can be a challenge, as it is clearly not practicable to use an independent assessor to analyse each location where work may take place as a user travels around with their portable.

5   One solution is to give portable DSE users sufficient training and information to make their own risk assessments and ensure that measures are taken to control risks (for example poor posture) whenever they set up their portable. This is discussed under advice to homeworkers; see paragraphs 27-29 of the main guidance. Portable users' risk assessments for, say, half an hour's work in a borrowed office can be quite informal and need not be written down. Where, however, a portable is in lengthy or repeated use in the same location, it would be appropriate for the user's risk assessment to be recorded, for example on a checklist. In all cases, portable users need to be alert to potential risks and report any problems to their employer.

6   As well as the risks common to both portables and desktop DSE work, the following additional risks may be associated specifically with portable DSE work and need to be taken into account by employers and users:

(a)   manual handling risks when moving between locations (bearing in mind that other equipment such as spare batteries, printers, or papers may add to the burden of the portable itself) (see *Getting to grips with manual handling*).[28]

(b)   Risk of theft possibly involving an assault.

Points to look for in choosing equipment and designing tasks to minimise risks are discussed in paragraphs 7-15 of this appendix.

Equipment, workstation and task requirements (regulation 3 and Schedule)

7   As with full-sized DSE, portables in prolonged use (and the workstations and working environments where they are used) are required to comply with the Schedule. The main difference is that the inherent requirements of portability may mean that some of the detailed requirements of paragraphs 2, 3 and 4 of the Schedule cannot be complied with in all respects. (This kind of non-compliance is allowed for in the circumstances described in the Schedule's paragraph 1.)

8   Users and employers should be aware that some design compromises inherent in portables can lead to postural or other problems (for example a bent neck, or headaches arising from the low, fixed position of the screen). One way of tackling such risks is to avoid prolonged use and take more frequent breaks. Another way, if working in an office, is to use the portable with a docking station; more advice on this is given in paragraph 11 of this appendix.

9   Some practical points to consider when **selecting portable computers** are as follows:

(a)   Look for as low a weight as possible (for example 3 kg or less) for the portable computer, and keep accessories as few and as light as possible.

(b)   Choose as large and clear a screen as possible, that can be used comfortably for the task to be done.

(c)   Where available, opt for a detachable or height-adjustable screen.

(d)   Specify as long a battery life as possible. Where practicable, provide extra transformer/cable sets so the user has a set in each main location where the portable is used, and only carries the computer, not the transformer/cables etc.

(e)   Give users a lightweight carrying case with handle and shoulder straps. To reduce risk of theft or assault, avoid manufacturer-branded laptop cases.

(f)   Look for tilt-adjustable keyboards on laptops.

(g)   Choose portables capable of being used with a docking station and/or with a facility for attaching an external mouse, keyboard and/or numeric keypad, where these are likely to help the user to work comfortably.

(h)   Check the portable has friction pads underneath to prevent it sliding across work surfaces when in use.

(i)   To cut working time and user stress, ensure the portable has sufficient memory and speed for the applications to be used.

(j)   For some tasks it may also be desirable to provide add-ons that improve usability and reduce maintenance time, such as (removable) CD-ROM drives and additional memory - but consider the weight penalty when deciding if this is appropriate.

(k)   For applications requiring use of a non-keyboard input device, opt for a portable with a touch pad, rollerball or external mouse rather than a 'nipple' trackpoint or isometric joystick device.

(l)   Many users find it more comfortable to use portables whose casing incorporates a space (wrist pad) between the keyboard and front edge.

10   Other points to consider when **planning tasks** involving portable computers are:

(a)   Think about weights to be carried. Where necessary (for example if workers are carrying substantial amounts of equipment and/or papers), carry out manual handling risk assessments with portable computer users.

(b)   Advise workers to set up their portable on a suitable worksurface wherever possible, and avoid use for extended periods in other situations. For example resting a portable on the user's lap is not only likely to induce a poor working posture but could result in discomfort due to the heat generated by the computer.

(c)   Provide docking stations or similar equipment (see paragraph 11 of this appendix) at workstations where portable computers will be in lengthy or repeated use.

(d)   Ensure that staff use portable computers only when away from their main place of work, or when docking station equipment is unavailable.

(e)   Minimise the use of portable computers in non-ideal locations such as motor vehicles.

(f)   Ensure that handheld computers for prolonged use are carefully selected for ergonomic features which match the requirements of the tasks undertaken. For example equipment to be used outdoors should be adequately waterproof, legible in bright sunlight, and keyboards and screens should be large enough to be used comfortably.

11   **Docking stations** are a way to avoid many of the ergonomic disadvantages of portables by allowing the use of a full-sized screen and/or keyboard (and mouse or other peripherals). Designs vary: some resemble a full-sized PC with a slot for the portable to be inserted; others comprise a screen, keyboard, mouse and/or other peripherals connected to the portable by cables or wireless links. There are also systems that provide a full-sized keyboard plus raiser blocks to enable the portable's own screen to be viewed at a more convenient height (see Figure 3). Height-adjustable stands for notebook computers are also available. In setting up any kind of docking station, the aim is for the user to achieve a comfortable working position allowing some variation in posture and having sufficient space for documents and anything else needed for their work tasks. The advice on workstations and working environments in Appendix 1 should be followed, treating the docking station in the same way as full-sized DSE.

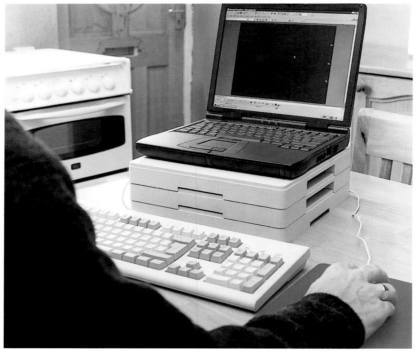

Figure 3

12   Risks of **theft and mugging** exist in some circumstances. They can be tackled by a combination of user training and task design; for example:

(a)   Do not design tasks in such a way that lone users are expected to carry or use portables in circumstances where theft is likely.

(b)   Tell all users to take sensible precautions such as not carrying portables in luggage with a computer manufacturer's branding; not leaving or using a portable in a parked car; and taking extra care in public places, or in other situations (or at times) where the risk of theft may be greater.

13   If the task involves risk from **manual handling**, employers and users can take commonsense steps to cut down the risk; for example:

(a)   Do not carry equipment or papers unless they are really likely to be needed.

(b)   Consider using a backpack to cut down strain on arms and distribute loads evenly across the body (or wheeled luggage might be worth considering).

(c)   Remember you may be able to avoid carrying heavy papers by sending them in advance, by post or e-mail, to your destination, or storing them electronically on the portable or on a disk.

The HSE guidance entitled *Getting to grips with manual handling*[28] gives more detailed advice on weights, precautions, etc.

Breaks or changes of activity (regulation 4)

14   Breaks or changes of activity are particularly important for portable users not working at a docking station. Such users need longer and more frequent breaks or changes of activity to compensate for poorer working environments, which can impact particularly on posture.

15   Employers whose staff use portables, particularly those who travel and work unsupervised, should remind them frequently of the need to take breaks. Break-monitoring software may be a useful aid (see paragraphs 65-67 of the main guidance for more detailed information on break-monitoring software).

Eyes and eyesight (regulation 5)

16   With regard to eyes and eyesight, there are few special considerations for portable users, although it may be helpful for the user to tell the optician doing any eye and eyesight test that a portable is used, as typical viewing distances may be somewhat shorter than for desktop DSE.

Training (regulation 6) and information (regulation 7)

17   Good health and safety training is particularly important for people who make any prolonged use of portables (including docking stations or handhelds.) Employers should ensure all such employees receive adequate training, including the following things specific to using a portable:

(a)   Advice on how to set up and use the equipment in the locations where it is to be used (bearing in mind the user needs sufficient knowledge of risks and precautions to, in effect, re-do the risk assessment whenever starting work in each location; as discussed in paragraphs 4-6 of this appendix).

(b)   Guidance on setting up and using a docking station, and additional precautions if using a portable computer when a docking station is not available (see Figure 4, which shows a setup that would not be acceptable for extended use).

Figure 4

(c)   Encouragement and advice on how to report promptly any symptoms of discomfort that may be associated with their use of portable DSE, and where to get further advice and help.

(d)   A reminder to take regular breaks, bearing in mind that increased DSE use is linked to an increasing risk of discomfort.

(e)   How to avoid unnecessary manual handling when carrying around portable DSE (and associated equipment and/or paperwork), and how to reduce risk from such manual handling as is unavoidable.

(f)   Advice on how to minimise risks from theft or mugging.

18   Managers of staff who use portable DSE should themselves receive health and safety training, so that they are aware of the issues and able and willing to take action to prevent health risks and respond to any problems reported. Key issues managers should be aware of are:

(a)   The need for regular breaks to avoid unnecessary use of DSE for extended periods.

(b)   Benefits of ensuring adequate variety in users' tasks.

(c)   Importance of health and safety training for users.

(d)   Reasons for providing docking station equipment wherever possible, and encouraging its use.

Work with a mouse, trackball or other pointing device

1   Most modern computers allow or require the use of other input devices besides the keyboard. Many such devices are available, the most common examples being the mouse and the trackball (or trackerball). This appendix refers to all these non-keyboard input devices as **pointing devices**, as they are commonly used to move the cursor around the screen in order to operate buttons, or to select and manipulate text, windows and other on-screen objects. Other pointing devices include joysticks, touchpads and touchscreens. Speech can be another form of non-keyboard input; though not strictly a pointing device, some brief guidance on speech interfaces is included below.

General considerations in choosing pointing devices

2   The majority of desktop computers are supplied with a computer mouse. In most situations the mouse will be an appropriate pointing device to use. However, there will be cases where an alternative device is more suitable or is preferred by the user. The mouse depends on having a suitable surface with enough space on which to use it. So, for example where space is very limited or where an individual has limited mobility in their arm, an alternative such as the trackball or touchpad may be more suitable. These alternative devices are frequently found on portable computers.

3   In choosing a pointing device there are a number of factors to consider:

(a)   **The environment in which it will be used.** Will the user be able to easily use the device at the workstation? Is there enough space? Can the user adopt a safe, comfortable working posture? Is there a suitable surface on which to use the device? Will other factors such as a dirty working environment or vibration affect its use?

(b)   **Individual characteristics.** Is the device the right size and shape for the user? Will right- and left-handed users be able to use it? Will the device be usable if the individual has any physical limitations (for example an existing upper limb disorder)?

(c)   **Task characteristics.** Does the task demand a lot of use to be made of the pointing device? Is a lot of fast and accurate positioning of the cursor required? Some devices are better than others in terms of speed and accuracy; for example the mouse tends to be more accurate but slightly slower than the trackball for fast, long cursor movements.

Using a pointing device

4   Many of the principles which apply to the setting up and use of a keyboard also apply to pointing devices. The following are particularly relevant:

(a)   **Positioning.** In general it is important to place the device so that it is fairly close to the midline of the user's body, not out to one side. The aim is to avoid the arm becoming stretched out from the shoulder as if reaching (see Figure 5). Aim to have the user's upper arm close to the side of their body with the elbow bent to approximately a right-angle (see Figure 6). The arm should feel reasonably relaxed. For devices such as the mouse which are used on a worksurface, the forearm or wrist can usefully be supported by the worksurface, or the elbow by the chair arm. If a keyboard is in use, the mouse needs to be positioned close to the

Figure 5

Figure 6

keyboard, on whichever side suits the user. Alternatively, if the keyboard is not being used, move it to one side and place the mouse closer to the centre. Similar advice applies to trackballs and joysticks, although in some cases these can be held in the user's other hand or supported on their lap.

(b) **The workstation**. The height of the worksurface is important. However, generally if the worksurface complies with the requirements of the Schedule then it should be suitable for use with a computer mouse. There may be some advantage in using wraparound worksurfaces (for example L-shaped desks) where a mouse is used. They may provide more support for the user's arm. However, such surfaces should not be so tightly curved that they restrict movement of users and their chairs.

(c) **Mousemats**. These are often helpful. They should have a smooth surface and be large enough to be suitable for the task. They should not have sharp edges which could put pressure on the soft tissue of the forearm or wrist. Special wrist rests are not a requirement. If used, they should be chosen with care to ensure they do not increase rather than decrease the risks.

(d) **Software settings**. It is important to configure the speed and sensitivity of the pointing device to suit the individual user. Standard office computer operating systems let the user adjust the gearing between the movement of the device and movement of the cursor, and the sensitivity of the buttons for double clicking.

(e) **Task organisation**. As with DSE work generally, periods of using a pointing device need to be interspersed with other activities. Intensive use of the device can often be reduced by training users to use alternative means of achieving the same ends, such as keyboard short cuts, especially for frequently performed functions (for example cut and paste). Users also need to be aware that it is better to remove their hand from the device when not actually using it (for example between bouts of intensive activity) so as to avoid prolonged static postures which can contribute to upper limb disorders. Periodically relaxing the arms by letting them hang down by the side of the body is a good idea.

(f) **Training**. This should not be neglected. Users may need to be taught how to set up and use their pointing devices, to get the best out of them and avoid risks.

(g) **Cleaning and maintenance**. This is needed for most pointing devices, and a system should be put in place for this. For example moving parts such as mouse balls and rollers may need periodic removal and/or cleaning. Such cleaning can be carried out by users but they may need training and reminding to do this.

(h) **Buying new equipment**. In purchasing new pointing devices, consider the device size, shape, handedness, number and position of buttons, ease of operation and user comfort.

Touchscreens

5   There are various uses for touchscreens. Some are built into the main display, others are stand-alone screens which have more in common with a keyboard or a bank of switches (for example a telephone switchboard). Some of the general advice given in paragraph 4 of this appendix on pointing devices applies also to touchscreens, but there are some special considerations:

(a) Touchscreens of both types need regular maintenance, particularly cleaning, if they are to be effective displays which comply with the Schedule. The screen's sensitivity to touch needs to be suitable, to ensure the screen is easy to use.

(b) Screen positioning may require some care. Applications using the main display as the touchscreen tend to be for use by the general public (for example automated bank teller machines, ticket machines). If used in the workplace it may be necessary to compromise between the optimal positions of the screen for viewing and for use as a touchscreen, dependent upon the extent of use for each activity.

(c)   Where the touchscreen is a dedicated additional display (for example for use as a telephone switchpanel) then it is also important that it is positioned where the user can both easily access it and see it. The angle at which it is set to the horizontal will be important if glare and reflections from ceiling lights are to be avoided. But the touchscreen also needs to be set so that the user can operate it without having to reach or adopt awkward postures. Advice should be sought from a suitable specialist (for example an ergonomist) if in any doubt how to achieve this.

(d)   Software needs to be suitably designed for the touchscreen. For example active areas should be big enough to respond equally accurately to users with large or small fingers. The layout of information on the screen also needs care as information at the bottom is often obscured by the user's hand.

Speech interfaces

6   Speech interfaces are becoming more readily available as the technology improves. Little is currently known about the health and safety aspects associated with their use, but some general guidance can be given. As with other forms of non-keyboard device, the characteristics of this form of input should be considered in assessing its suitability for the task, the environment and the user. Relevant factors include the position of the microphone and how it is supported; the software settings; how noisy the environment is in which it will be used; whether use will prove to be a distraction to others; and possible effects on the individual user, for example, voice strain from having to adapt their mode of speech to suit the requirements of the interface. Voice recognition software is becoming more powerful and versatile and can be used exclusively or in conjunction with other input devices. If used, particular attention should be given to initial and refresher training of users and specialist IT support for them.

VDU WORKSTATION CHECKLIST

Workstation location and number (if applicable): ....

User: ....

Checklist completed by: ....

Assessment checked by: ....

Date of assessment: ....

Any further action needed?   YES/NO

Follow-up action completed on: ....

This checklist can be used as an aid to risk assessment and to help comply with the Schedule to the Health and Safety (Display Screen Equipment) Regulations.

Work through the checklist, ticking either the 'yes' or 'no' column against each risk factor:

- 'Yes' answers require no further action.
- 'No' answers will require investigation and/or remedial action by the workstation assessor. They should record their decisions in the 'Action to take' column. Assessors should check later that actions have been taken and have resolved the problem.

Remember the checklist only covers the workstation and work environment. You also need to make sure that risks from other aspects of the work are avoided, for example by giving users health and safety training, and providing for breaks or changes of activity. Advice on these is given in the main text of the guidance.

VDU workstation checklist

RISK FACTORS	Tick answer		THINGS TO CONSIDER	ACTION TO TAKE
	YES	NO		

1 Display screens

RISK FACTORS	YES	NO	THINGS TO CONSIDER	ACTION TO TAKE
Are the characters clear and readable? **Health and safety** ✓ **Health and safety** ✗			Make sure the screen is clean and cleaning materials are made available. Check that text and background colours work well together.	
Is the text size comfortable to read?			Software settings may need adjusting to change text size.	
Is the image stable, ie free of flicker and jitter?			Try using different screen colours to reduce flicker, eg darker background and lighter text. If problems still exist, get the set-up checked, eg by the equipment supplier.	
Is the screen's specification suitable for its intended use?			For example, intensive graphic work or work requiring fine attention to small details may require large display screens.	
Are the brightness and/or contrast adjustable?			Separate adjustment controls are not essential, provided the user can read the screen easily at all times.	
Does the screen swivel and tilt?			Swivel and tilt need not be built in; you can add a swivel and tilt mechanism. However, you may need to replace the screen if: ● swivel/tilt is absent or unsatisfactory; ● work is intensive; and/or ● the user has problems getting the screen to a comfortable position.	
Is the screen free from glare and reflections?			Use a mirror placed in front of the screen to check where reflections are coming from. You might need to move the screen or even the desk and/or shield the screen from the source of reflections. Screens that use dark characters on a light background are less prone to glare and reflections.	
Are adjustable window coverings provided and in adequate condition?			Check that blinds work. Blinds with vertical slats can be more suitable than horizontal ones. If these measures do not work, consider anti-glare screen filters as a last resort and seek specialist help.	

RISK FACTORS	Tick answer		THINGS TO CONSIDER	ACTION TO TAKE
	YES	NO		

2 Keyboards

RISK FACTORS	YES	NO	THINGS TO CONSIDER	ACTION TO TAKE
Is the keyboard separate from the screen?			This is a requirement, unless the task makes it impracticable (eg where there is a need to use a portable).	
Does the keyboard tilt?			Tilt need not be built in.	
Is it possible to find a comfortable keying position?			Try pushing the display screen further back to create more room for the keyboard, hands and wrists. Users of thick, raised keyboards may need a wrist rest.	
Does the user have good keyboard technique?			Training can be used to prevent: ● hands bent up at wrist; ● hitting the keys too hard; ● overstretching the fingers.	
Are the characters on the keys easily readable?			Keyboards should be kept clean. If characters still can't be read, the keyboard may need modifying or replacing. Use a keyboard with a matt finish to reduce glare and/or reflection.	

3 Mouse, trackball etc

RISK FACTORS	YES	NO	THINGS TO CONSIDER	ACTION TO TAKE
Is the device suitable for the tasks it is used for?			If the user is having problems, try a different device. The mouse and trackball are general-purpose devices suitable for many tasks, and available in a variety of shapes and sizes. Alternative devices such as touchscreens may be better for some tasks (but can be worse for others).	
Is the device positioned close to the user?			Most devices are best placed as close as possible, eg right beside the keyboard. Training may be needed to: ● prevent arm overreaching; ● tell users not to leave their hand on the device when it is not being used; ● encourage a relaxed arm and straight wrist.	

63

RISK FACTORS	Tick answer		THINGS TO CONSIDER	ACTION TO TAKE
	YES	NO		
Is there support for the device user's wrist and forearm?			Support can be gained from, for example, the desk surface or arm of a chair. If not, a separate supporting device may help. The user should be able to find a comfortable working position with the device.	
Does the device work smoothly at a speed that suits the user?			See if cleaning is required (eg of mouse ball and rollers). Check the work surface is suitable. A mouse mat may be needed.	
Can the user easily adjust software settings for speed and accuracy of pointer?			Users may need training in how to adjust device settings.	

4 Software

Is the software suitable for the task?			Software should help the user carry out the task, minimise stress and be user-friendly. Check users have had appropriate training in using the software. Software should respond quickly and clearly to user input, with adequate feedback, such as clear help messages.	

5 Furniture

Is the work surface large enough for all the necessary equipment, papers etc?			Create more room by moving printers, reference materials etc elsewhere. If necessary, consider providing new power and telecoms sockets, so equipment can be moved. There should be some scope for flexible rearrangement.	
Can the user comfortably reach all the equipment and papers they need to use?			Rearrange equipment, papers etc to bring frequently used things within easy reach. A document holder may be needed, positioned to minimise uncomfortable head and eye movements.	
Are surfaces free from glare and reflection?			Consider mats or blotters to reduce reflections and glare.	

RISK FACTORS	Tick answer		THINGS TO CONSIDER	ACTION TO TAKE
	YES	NO		
Is the chair suitable? Is the chair stable? Does the chair have a working: ● seat back height and tilt adjustment? ● seat height adjustment? ● swivel mechanism? ● castors or glides?			The chair may need repairing or replacing if the user is uncomfortable, or cannot use the adjustment mechanisms.	
Is the chair adjusted correctly? 			The user should be able to carry out their work sitting comfortably. Consider training the user in how to adopt suitable postures while working. The arms of chairs can stop the user getting close enough to use the equipment comfortably. Move any obstructions from under the desk.	
Is the small of the back supported by the chair's backrest?			The user should have a straight back, supported by the chair, with relaxed shoulders.	
Are forearms horizontal and eyes at roughly the same height as the top of the VDU?			Adjust the chair height to get the user's arms in the right position, then adjust the VDU height, if necessary.	
Are feet flat on the floor, without too much pressure from the seat on the backs of the legs?			If not, a foot rest may be needed.	

RISK FACTORS	Tick answer		THINGS TO CONSIDER	ACTION TO TAKE
	YES	NO		

6 Environment

RISK FACTORS	YES	NO	THINGS TO CONSIDER	ACTION TO TAKE
Is there enough room to change position and vary movement?			Space is needed to move, stretch and fidget. Consider reorganising the office layout and check for obstructions. Cables should be tidy and not a trip or snag hazard.	
Is the lighting suitable, eg not too bright or too dim to work comfortably?			Users should be able to control light levels, eg by adjusting window blinds or light switches. Consider shading or repositioning light sources or providing local lighting, eg desk lamps (but make sure lights don't cause glare by reflecting off walls or other surfaces).	
Does the air feel comfortable?			VDUs and other equipment may dry the air. Circulate fresh air if possible. Plants may help. Consider a humidifier if discomfort is severe.	
Are levels of heat comfortable?			Can heating be better controlled? More ventilation or air-conditioning may be required if there is a lot of electronic equipment in the room. Or, can users be moved away from the heat source?	
Are levels of noise comfortable?			Consider moving sources of noise, eg printers, away from the user. If not, consider soundproofing.	

7 Final questions to users...

- Ask if the checklist has covered all the problems they may have working with their VDU.

- Ask if they have experienced any discomfort or other symptoms which they attribute to working with their VDU.

- Ask if the user has been advised of their entitlement to eye and eyesight testing.

- Ask if the user takes regular breaks working away from VDUs.

Write the details of any problems here:

Sources of information and advice

The following gives a list of useful sources of information and advice

Trade union guidance

The TUC and several trade unions have produced detailed publications on a wide variety of display screen work, in some cases specific to particular occupations.

Sources of expertise

Ergonomics Society
Devonshire House, Devonshire Square, Loughborough LE11 3DW
Tel: 01509 234904
Web address: www.ergonomics.org.uk

The College of Optometrists
42 Craven St, London WC2N 5NG
Tel: 020 7839 6000.
Web address: www.college-optometrists.org
The College is a registered charity and is the examining and professional body for optometrists. It has issued a statement of good practice concerning eye tests for DSE workers. This can be found under 'Work with display screen equipment' in the College's guidance for professional conduct, on their website.

The Association of Optometrists
61 Southwark Street, London SE1 0HL
Tel: 020 7261 9661
Web address: www.assoc-optometrists.org
The Association has issued brief guidance on visual standards for VDU use, and a sample report form to be filled in by optometrists.

Chartered Institution of Building Services Engineers (CIBSE)
222 Balham High Road, London SW12 9BS
Tel: 020 8675 5211
Web address: www.cibse.org

HSE's Employment Medical Advisory Service
HSE regional offices (listed in your local telephone directory under 'Health and Safety Executive').

Epilepsy Action
Freephone helpline: 0808 800 5050

Repetitive Strain Injury Association
Freephone RSI helpline: 0800 018 5012
Web address: www.rsi.org.uk

BackCare
16 Elmtree Road, Teddington, Middlesex TW11 8ST
Tel: 020 8977 5474.
Web address: www.backpain.org
BackCare is a charity for healthier backs.

AbilityNet
Tel: 0800 269545
Web address: www.abilitynet.org.uk
AbilityNet is a national charity and a provider of expertise on computing and disability.

BECTA
Web address: www.becta.org.uk
BECTA is the Government's lead agency for ICT in education.

References and further reading

References

1   *The law on VDUs: An easy guide: Making sure your office complies with the Health and Safety (Display Screen Equipment Regulations) 1992 (as amended in 2002)* HSG90 HSE Books 2003 ISBN 0 7176 2602 4

2   *Health and Safety at Work etc Act 1974 Ch 37* The Stationery Office 1974 ISBN 0 10 543774 3

3   *Management of Health and Safety at Work. Management of Health and Safety at Work Regulations 1999. Approved Code of Practice and guidance* L21 (Second edition) HSE Books 2000 ISBN 0 7176 2488 9

4   *Workplace health, safety and welfare. Workplace (Health, Safety and Welfare) Regulations 1992. Approved Code of Practice* L24 HSE Books 1992 ISBN 0 7176 0413 6

5   *Safe use of work equipment. Provision and Use of Work Equipment Regulations 1998. Approved Code of Practice and guidance* L22 (Second edition) HSE Books 1998 ISBN 0 7176 1626 6

6   *Essentials of health and safety at work* HSE Books 1994 ISBN 0 7176 0716 X

7   *Understanding ergonomics at work: Reduce accidents and ill health and increase productivity by fitting the task to the worker* INDG90 (rev 2) 2003 (single copy free or priced packs of 15 ISBN 0 7176 2559 0)

8   *Homeworking: Guidance for employers and employees on health and safety* INDG226 HSE Books 1996 (single copy free or priced packs of 15 ISBN 0 7176 1204 X)

9   BS EN ISO 9241 *Ergonomics requirements for office work with visual display terminals (VDTs)* (Parts 1 to 17, 1992 to 2000) British Standards Institution

10 BS 3044: 1990 *Guide to ergonomic principles in the design and selection of office furniture* British Standards Institution

11 BS EN 527: Part 1: 2000 *Office furniture. Work tables and desks. Dimensions* British Standards Institution

12 BS EN 1335: Part 1: 2000 *Office furniture. Office work chair. Dimensions. Determination of dimensions* British Standards Institution

13 BS EN ISO 11064 *Ergonomic design of control centres* (Parts 1 to 7, 2000 ongoing) British Standards Institution

14 BS EN 614 *Safety of machinery. Ergonomic design principles* (Parts 1 and 2, 1995 and 2000) British Standards Institution

15 BS EN ISO 13407: 2000 *Human-centred design processes for interactive systems* British Standards Institution

16 BS EN 61508 *Functional safety of electrical/electronic/programmable electronic safety-related systems* (Parts 1-7, 2002) British Standards Institution (also known as IEC 61508)

17   *Seating at Work* HSG57 (Second edition) HSE Books 1997
      ISBN 0 7176 1231 7

18   *Lighting at Work* HSG38 (Second edition) HSE Books 1997
      ISBN 0 7176 1232 5

19   *Lighting guide: The visual environment for display screen use* LG3 Chartered
      Institute of Building Services Engineers 1996 ISBN 0 900953 71 3

20   J M Harrington, J T Carter, L Birrell, D Gompertz. Surveillance case
      definitions for work related upper limb pain syndromes *Occupational and
      Environmental Medicine* 1998 55(4) 264-271

21   *Thermal comfort in the workplace: Guidance for employers* HSG194 HSE
      Books 1999 ISBN 0 7176 2468 4

22   *Back in work: Managing back pain in the workplace - A leaflet for employers
      and workers in small businesses* INDG333 HSE Books 2000 (single copy free)

23   *Work-related stress: A short guide* Leaflet INDG281(rev1) HSE Books
      2002 (single copy free or priced packs of 10 ISBN 0 7176 2112 X)

24   *Tackling work related stress: A manager's guide to improving and maintaining
      employee health and well-being* HSG218 HSE Books 2001
      ISBN 0 7176 2050 6

25   *Health effects related to the use of visual display units* National Radiological
      Protection Board. Documents of the NRPB 1994

26   *New and expectant mothers at work: A guide for employers* HSG122 (Second
      edition) HSE Books 2002 ISBN 0 7176 2583 4

27   *Upper limb disorders in the workplace* HSG60 (Second edition)
      HSE Books 2002 ISBN 0 7176 1978 8

28   *Getting to grips with manual handling: A short guide for employers*
      INDG143(rev1) (single copy free or priced packs of 15
      ISBN 0 7176 1754 8)

Further reading

*Board statement on restrictions on human exposure to static and time varying
electromagnetic fields and radiation* National Radiological Protection Board.
Documents of the NRPB 1993

Checkouts and musculoskeletal disorders Leaflet INDG269 HSE Books 1998
(single copy free or priced packs of 15 ISBN 0 7176 1539 1)

The Health and Safety (Display Screen Equipment) Regulations 1992
SI 1992/2792 The Stationery Office 1992 ISBN 0 11 025919 X

The Health and Safety (Miscellaneous Amendments) Regulations 2002
SI 2002/2174 The Stationery Office 2002 ISBN 0 11 042693 2

Reducing error and influencing behaviour HSG48 (Second edition) HSE Books
1999 ISBN 0 7176 2452 8

Working with VDUs Leaflet INDG36(rev1) HSE Books 1998 (single copy free
or priced packs of 10 ISBN 0 7176 1504 9)

The Stationery Office (formerly HMSO) publications are available from
The Publications Centre, PO Box 276, London SW8 5DT
Tel: 0870 600 5522 Fax: 0870 600 5533 Website: www.tso.co.uk
(They are also available from bookshops.)

British Standards are available from BSI Customer Services,
389 Chiswick High Road, London W4 4AL Tel: 020 8996 9001
Fax: 020 8996 7001 Website: www.bsi-global.com

Printed and published by the Health and Safety Executive   2/03   C100